D1225157

THE ATLAS OF MERCURY

CHARLES A. CROSS & PATRICK MOORE

FOREWORD BY SIR BERNARD LOVELL

MITCHELL BEAZLEY PUBLISHERS LIMITED

Foreword

The decade 1960 to 1970 will be remembered in history as the epoch when man reached the Moon and returned safely to Earth, but astronomers may well reflect on the period as one during which scientific instruments were first carried to the vicinity of the planets by space probes. It was from those less publicized activities that a great advance in our knowledge of the Solar System occurred. The revolution of the sixteenth and seventeenth centuries associated with the work of Copernicus, Galileo and Kepler marked a complete change in our dynamical view of the Solar System. The abandonment of the age-long belief that the Earth was fixed in space, the realization by Kepler that the planets moved in elliptical orbits around the Sun and the gravitational theory of Newton established a satisfactory model for the motions of the planets which has required only relatively minor modifications in the succeeding three centuries.

Throughout this period our knowledge of the physical condition of the planets has advanced very slowly. The difficulty has been that although the great optical telescopes built on Earth have been able to penetrate far into space they have had only marginal advantages for the study of the planets. For example, on the occasional close approaches to Earth of the planet Mars the best definition of features on the surface as seen through a telescope on Earth is only about 30 to 45 miles (50 to 70 km). The surface of Venus never can be seen from Earth because it is perpetually covered in cloud, and although the observations of Mercury do not suffer that problem there is an additional difficulty that because of the nature of its orbit close to the Sun the planet can never be viewed from Earth against a completely dark sky. The far greater distances of the outer planets, Jupiter, Saturn, Uranus, Neptune and Pluto, increase the difficulties of observation from Earth. The net result of observations of the planets made during the last century with the best available instruments has left major uncertainties about the nature of their surfaces, the constitution of their atmospheres and—in the case of Venus and Mercury—even the speed at which they are rotating.

The first successful planetary probe was the American Mariner 4, launched in 1964, passing Mars at a distance of 6,100 miles (9,800 km) in July 1965 and transmitting to Earth photographs which revealed that the surface was heavily cratered. Since that time the American spacecraft to the planet, culminating in the landing of the Viking spacecraft in 1976, have transmitted to Earth photographs and other data which have falsified nearly every previous text on the planet. Even more cataclysmic changes have been wrought by the succession of Soviet spacecraft to Venus, first penetrating the atmosphere in 1967 and recording a hot, poisonous and hostile climate in strange contrast to the beauty of the planet as seen from Earth. Mercury presented a far more difficult problem for space technology because of the high energy needed to place a probe in the vicinity of this planet. However, in recent years techniques have developed to the extent that the problem has been solved by making use of the gravitational interaction of the probe with the planet Venus. The American Mariner 10 spacecraft, launched on 3 November, 1973, made a fly-by of Venus at a distance of 3,570 miles (5,750 km) on 5 February, 1974. Assisted by the gravitational attraction of Venus the spacecraft then entered an orbit which brought it to within 470 miles (750 km) of Mercury on 29 March, 1974. Two other close approaches of Mariner 10 with Mercury occurred on 21 September, 1974, and on 16 March, 1975. Many of the photographs transmitted to Earth have resolutions of about 110 yards (100 m) and a surface, only previously seen indistinctly by terrestrial observers, has been revealed in remarkable detail. A map of Mercury compiled from the best evidence available before the flight of Mariner is seen to bear little resemblance to the lunar-like surface of craters, scarps, ridges and plains now revealed.

It is admirable that this almost complete reorientation of our ideas about Mercury is now presented in this book. Apart from the surface features we now know about the tenuous atmosphere, the magnetic field—and that the age-old belief that Mercury rotates so that one face is always presented to the Sun is quite wrong. Of course, future spacecraft will still further increase our knowledge of the planet, but it is unlikely that a probe will be placed in orbit around Mercury for at least ten years—and until that occurs this text is likely to remain the most concise and accurate account of the planet available.

Professor Sir Bernard Lovell, OBE, LLD, DSc, FRS

©Mitchell Beazley Publishers Limited 1977
First edition published 1977
Reproduction of any kind in part or in whole
in any country throughout the world is reserved
by Mitchell Beazley Publishers Limited
ISBN 0 85533 115 1
Mitchell Beazley Publishers Limited
87–89 Shaftesbury Avenue, London W1V 7AD
Photoset in Great Britain by Tradespools Ltd, Frome
Printed in Great Britain by
Sir Joseph Causton & Sons Ltd,
London and Eastleigh

Editor Lawrence Clarke
Art Editor Martin Bronkhorst
Asst. Art Editor Allison Blythe
Picture Researcher Susan Pinkus
Illustrators Marilyn Bruce, Mike Saunders, Alan Suttie

Publisher Bruce Marshall
Art Director John Bigg
Executive Editor Paul Bradwell
Production Director Michael Powell

Contents

Introduction

Now, for the first time, following the flight of Mariner 10, it is possible to give a comprehensive survey of the Mercurian scene, and to interpret the features that have been recorded there. In any scientific study an understanding of the gradual development of our knowledge is important, and this is why this atlas deals not only with the space-probe results but also with the patient work of earlier observers using telescopes based upon the surface of the Earth. These observers had a difficult task. Mercury is small, and never comes close to us, so there is little to be seen upon it even under the best conditions. Men such as Schiaparelli in Italy and, above all, Antoniadi in France worked patiently and efficiently; the fact that their charts and many of their conclusions bore little resemblance to the Mariner results was no fault of theirs. Their failure is merely an indication of the difficulty of terrestrial observations even with the largest and most powerful of instruments. Nothing more could be achieved until the development of space research methods.

The background to the pioneer space-probe Mariner 10—its construction and journey—is described and in part explains the extent of the photographic coverage and the method of mapping arising from it. Thus the main aim of the atlas has been to select the very best of the Mariner photographs, compile a new chart from them, and then reproduce some of the photographs themselves, choosing those which bring out the most spectacular and the most important characteristics. In addition to the photographic coverage other experiments carried out by Mariner—the method of analysis and the results—are discussed.

Mariner 10 made three active passes of Mercury transmitting data of high quality, including the unexpected discoveries of a residual atmosphere and a magnetic field. The photographic results sent back have provided all our present knowledge of the surface of the planet. There are craters similar in type to those of the Moon as well as many other features—some of them unlike anything else so far found in the Solar System. The Mariner photographs and their interpretation, therefore, provide the only means of undertaking close studies of Mercury upon which theories, encompassing the whole field of planetary origin and evolution, may be based.

The innermost planet

Two Mercurys— one at dusk, another at dawn

To observers of the ancient world, only five of the nine planets, apart from the Earth—which make up the Solar System—were known to exist. Of these, Venus, Mars, Jupiter and Saturn were easy to recognize, because all are brilliant naked-eye objects. The fifth, Mercury—the closest planet to the Sun—was something of a mystery. It is visible with the naked eye only when low down on the horizon—in the west just after sunset and in the east just before sunrise— and it is never seen against a dark sky. It was some time before the old stargazers realized that "evening Mercury" and "morning Mercury" were one and the same object.

The difficulty of naked-eye observation stems from the fact that Mercury seems to keep relatively close to the Sun. Even at its greatest elongation, the angular separation between Mercury and the Sun is only 27°45', which is not very much. Yet Mercury is not faint; at its brightest it may outshine any star in the sky, apart from Sirius. It appears elusive only because it is always seen against a bright sky.

The significance of Mercury before the invention of the telescope

The name we now give to the planet is that of the quick-moving messenger of the gods—to the Ancient Greeks, Hermes, and to the Romans, Mercury. Yet it seems that the planet had been identified well before the Greeks named it. The Egyptians certainly knew it well; it was from Egypt that the first recorded observation was made in 265 BC. Later astrologers naturally paid great attention to it, and its significance is reflected in its association with the day of the week Wednesday: *Mercurii dies*, or Mercury's day, in Latin; *mercredi* in French. The official symbol, still sometimes used, is ☿ .

By the end of the Classical period the movements of Mercury were well known (even though ancient astronomers thought that it moved around the Earth, like the other planets, rather than around the Sun) and its status as a planet was well established. Before the telescopic era nothing was known about Mercury's physical appearance, since its phases, or apparent changes of shape from crescent to full, are quite beyond the range of naked-eye observation.

Modern observers and the problem of rotation

Telescopes were invented in the seventeenth century, but these early, low-powered instruments—known as refractors, and using glass lenses to collect their light—revealed few of Mercury's secrets. The first observer to note the phases seems to have been Johannes Hevelius (1611–87)— best remembered today as being the compiler of a map of the Moon in 1645. In the following century Mercury was closely studied by Johann Hieronymus Schröter, also a great observer of the Moon; but even Schröter could see virtually nothing on Mercury's tiny disk.

A new series of observations was begun in 1881 by G. V. Schiaparelli, using 8½-in (22-cm) and 19-in (49-cm) refractors at Milan. Schiaparelli compiled a map of the surface, showing bright and dark features, and concluded that Mercury must have a synchronous rotation; if so, it would keep the same hemisphere turned to the Sun all the time, so that the rotation period of the planet would be the same as its orbital period around the Sun (approximately 88 Earth-days). He also followed the practice of observing Mercury when both it and the Sun were high above the horizon.

Later observers also drew bright and dark features— notably E. M. Antoniadi (1870–1943), who was responsible for much of the basic work concerning the planet as well as the nomenclature incorporated on more modern maps.

We now know that the rotation period is not the same as the orbital period; but it is true that when Mercury is best placed for observation the same face is always presented to us. This may well have been what misled earlier observers. They were wedded to the idea of a synchronous rotation, and so their interpretations were entirely wrong; but it is not impossible that some of the features they recorded are genuine.

The Astrologers' Mercury
This medieval woodcut, published in Germany in 1498, depicts Mercury as the bringer of misfortune—the accepted interpretation of astrologers throughout medieval Europe. This was in marked contrast to the Babylonian view of Mercury as the bearer of riches, a belief later adopted by the Arabs. Such attributes bore little relationship to observational work. It was the Roman Mercury that reflected the planet's most obvious characteristic—its quick movement as seen with the naked eye. It is not surprising that the planet assumed the name of the fleet-footed messenger.

High mountains, dark patches
Johann Schröter (1745–1816), one of the foremost planetary astronomers of his day, observed Mercury from the observatory at Lilienthal in Germany between 1779 and 1813. He published drawings showing features that he believed to be high mountains, estimating the height of one of them as 12 miles (20 km). He also drew dark patches, from which other astronomers derived a rotation period for the planet of 24 hours 0 minutes 53 seconds. Three typical drawings are shown; the apparent blunting of the horns, or cusps, is a feature first detected by Schröter, but is now accepted as an optical effect.

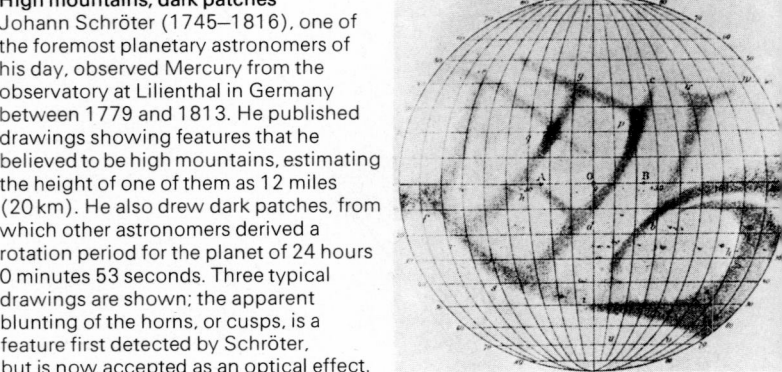

Schiaparelli's observations
The Italian Giovanni Schiaparelli (1835–1910) drew this map of Mercury based on 150 drawings of observations at the Brera Observatory between 1881 and 1889. He stated that the patches "almost always showed up in the form of extremely delicate, dark streaks, which under normal observing conditions could be recognized only with much effort and great consideration . . . all these streaks are clear brown in colour against a rosy background, always smoky, not easily seen against the background, and difficult to distinguish. Not surprisingly these features were later shown to be optical illusions.

Canals on Mercury?

Percival Lowell (1855–1916) made many drawings of Mercury, and recorded linear features similar to the canal network (which he believed to be artificial) on Mars. However, his description of Mercury emphasizes its natural appearance: "Its lines—more difficult than the canals of Mars, for we see Mercury four times as far off when best placed as we do Mars—though roughly linear, are not unnatural in appearance even at that great distance, and show irregularities suggestive of cracks." From his drawings he produced a planisphere, which bears no resemblance to other contemporary maps.

High-altitude refractor photography

French observers, notably B. Lyot and H. Camichel, concentrated on photographing surface detail on Mercury from the Pic du Midi Observatory in the Pyrenees, which is sited at an altitude of 9,000 ft (2,750 m) above sea-level, where atmospheric conditions are extremely good; the 23½-in (60-cm) refractor, specially adapted to planetary work, was the principal telescope used. The planisphere shown was drawn in 1942 from the results of their photographs. Lyot concluded that the albedo of Mercury was 13 per cent, slightly less than the Moon, and that the surface layers were lunar-like.

Optical drawbacks

Werner Sandner, a German observer, compiled his map in the 1950s, mainly from observations made with a refractor that was significantly smaller than that used at the Pic du Midi Observatory. Although there is some similarity to Schiaparelli's map in certain regions, most of the features illustrated are the result of misleading optical effects—as is inevitable when the observer is straining to catch fine details that are in fact beyond the resolution of his telescope. It is now accepted that small telescopes tend to make elusive features appear narrower and sharper than they really are.

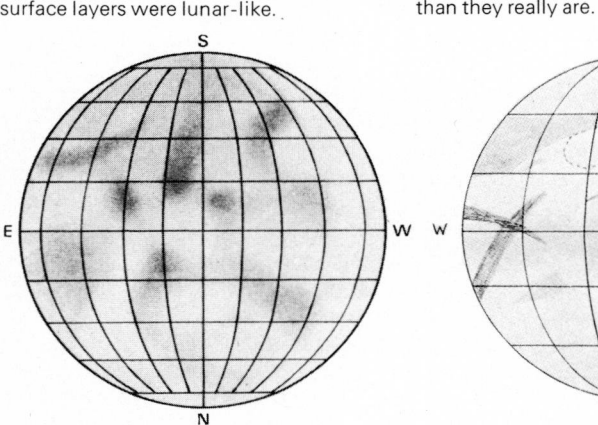

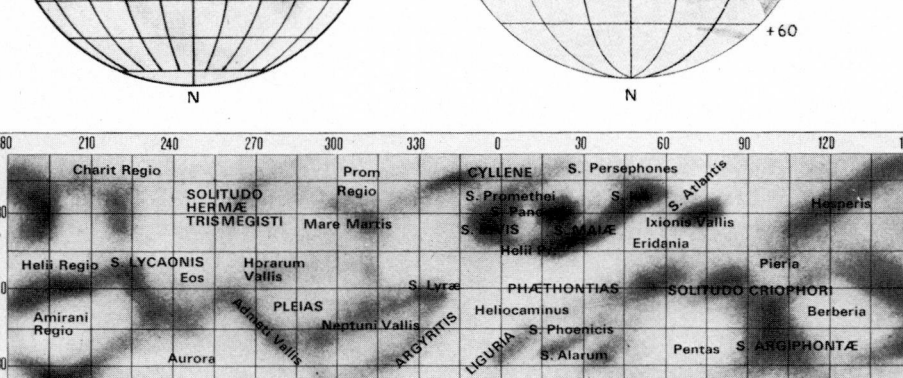

The 25-hour rotation theory

W. F. Denning (1848–1931) was one of the leading English observers of Mercury, although he is better remembered for his pioneer work in the study of meteors. At Bristol, Denning used a refractor with an aperture of 9½ in (24 cm). He recorded that the patches on Mercury were easily visible, and "so pronounced that they suggest an analogy with those of Mars"; from them he derived a rotation period of 25 hours. The four drawings were made in 1882. There is no correlation with the Mariner results, and E. M. Antoniadi was mistaken in writing that Denning was "the first to show genuine patches".

First hints of Mercurian atmosphere

H. Camichel and A. Dollfus produced a new map of Mercury in the 1950s which included all the features then regarded as being well defined (upper). The nomeclature used was that of Antoniadi, though somewhat extended and modified. There is some agreement with Antoniadi's own map and with that of C. Chapman (lower). In 1953 Dollfus reported that he had detected signs of a very tenuous atmosphere. Surprisingly, considering the similarities of surface detail between the Moon and Mercury, Mariner 10 did record an atmosphere, although it is considerably more tenuous than Dollfus' estimate.

The definitive map until Mariner

C. Chapman in the United States compiled the last pre-Mariner map of Mercury (above) from a selection of 130 drawings and photographs made by observers all over the world. He again incorporated Antoniadi's nomenclature. Chapman was able to draw upon the best available evidence, yet the result bears only a slight resemblance to contemporary attempts and little correlation to the Mariner views. The difficulty of observing Mercury from Earth is clearly emphasized. Even the largest telescopes cannot show Mercury as well as the Moon can be seen with the naked eye.

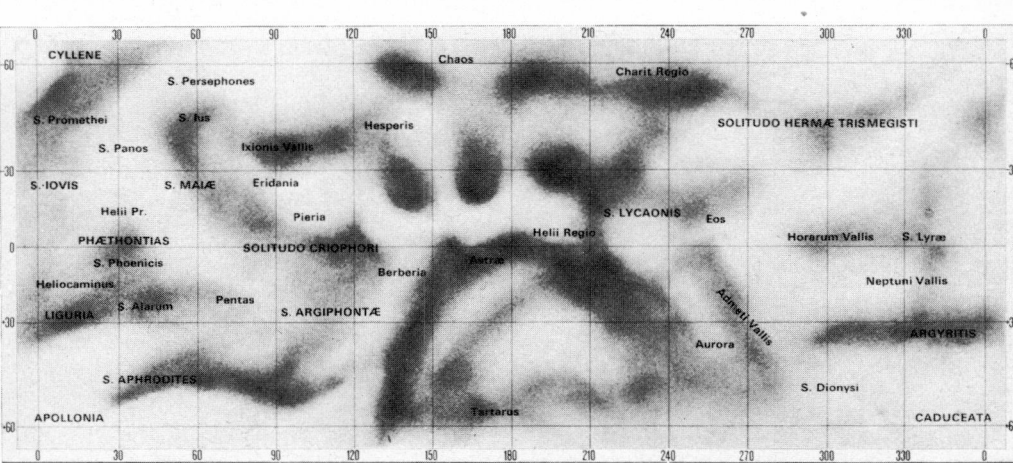

Antoniadi and "La Planète Mercure"

A foundation for all pre-Mariner Mercury work

Until the Mariner 10 probe began sending back data on Mercury in 1974, revealing the extent of ignorance and speculation about the nature of the planet and reversing many of the theories until then popularly held, astronomers had relied largely on the pioneering work of E. M. Antoniadi. From a long series of observations made during the late 1920s he compiled a map which was as good as any subsequently drawn up from Earth-based data, and his Mercurian nomenclature was incorporated in all serious attempts at charting the surface before Mariner. Not that Antoniadi was totally wrong: he was correct in his recording of the planet's phases, in his interpretation of observed luminous spots and in the phenomenon of the "black drop". His fundamental error was in assuming the planet had a synchronous rotation. Antoniadi published his findings in *La Planète Mercure* in 1934, which as an historical record clearly illustrates the limitations of all observations made from Earth of the more elusive celestial objects. Nevertheless, the work remains an outstanding example of reasoned scientific deduction based on methodical observation.

Antoniadi's method of observing and recording

A Greek-born astronomer who spent much of his life in France, Antoniadi began his serious observing work as assistant to the famous astronomer Camille Flammarion. He then became associated with the Paris Observatory, and made extensive use of the great 32½-in (83-cm) refractor at Meudon, one of the best telescopes in the world for planetary research. With it, Antoniadi made his classic observations of Mars, and concentrated on Mercury.

Like Schiaparelli, Antoniadi made virtually all his observations in broad daylight, with both Mercury and the Sun high above the horizon. His general comments are interesting: "On my drawings I have recorded only the patches which were seen with certainty. . . . In general the patches were very pale and difficult to distinguish, but they were genuine, and their colour seemed greyish, similar to that of the lunar seas."

The names he gave to the various patches were derived from Greek and Egyptian mythology—for instance, he called one shading the "Solitudo Hermae Trismegisti", or the "Wilderness of Hermes the Thrice Greatest". He considered it very likely that these darkish and brighter patches were of the same basic nature as the plains and uplands of the Moon.

His descriptions of such features were precise. Thus the so-called "Solitudo Iovis" was a "rounded patch, very dark; dimensions nearly equal to those of France. . . . It was nearly always easy to see in the great refractor, and very dark when it was not enfeebled by local veils."

The belief in a Mercurian atmosphere

The reference to "local veils" is highly significant. Antoniadi asserted that the density of the Mercurian atmosphere was just sufficient to support "excessively fine dusty particles" but was in no way comparable with the Earth's atmosphere. Yet, surprisingly, he recorded that the localized obscurations on Mercury were "more frequent and more obliterating than those of Mars". He also strongly supported Schiaparelli's theory that the rotation of Mercury was synchronous.

It was only with the development of space-probes that Antoniadi's work on Mars and Mercury could be properly checked. With Mars he proved to be remarkably accurate, and in the main his nomenclature has survived. With Mercury there is very little correlation between his map and modern charts based upon the Mariner 10 photography. Equally, although his assumption of the existence of an atmosphere was correct, his reasoning behind that assumption was not; the atmosphere of the planet is far too tenuous to hold any material in suspension. Finally, we now know that the rotation period is not synchronous, though it is true that when Mercury is best placed for observation from Earth the same regions are turned towards us.

The great "atmospheric veil"
These observations were made with the 32½-in (83-cm) Meudon refractor from 1927 to 1929. Antoniadi was very much concerned with his so-called "whitish veils". Thus on drawings A and B he recorded what he termed "a great atmospheric veil extending 3,000 km along the eastern limb of Mercury", whereas in the last drawing, C, he described the southern markings as "very intense". The drawings clearly show the extreme difficulty of seeing any markings on Mercury which are sufficiently definite for their positions to be charted, and this is no doubt why Antoniadi was misled into believing in local veilings.

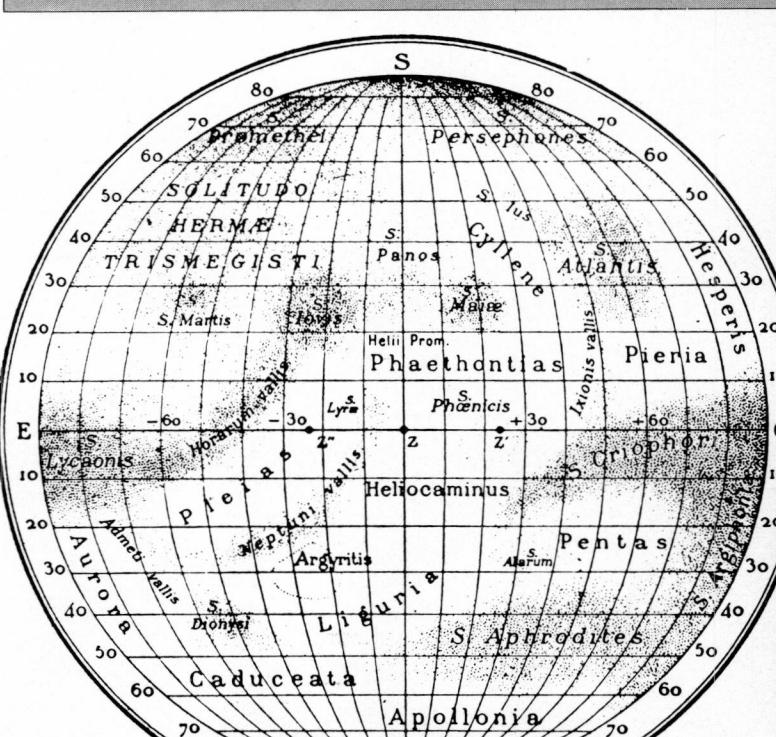

Light and shade from Meudon
Antoniadi's chart of Mercury was compiled from results based exclusively on the Meudon 32½-in (83-cm) refractor; most of his main observations were made between 1924 and 1929. In general, he recorded only features that he had seen on more than one occasion, and which he regarded as being definitely real. They consisted of dark patches (such as Solitudo Hermae Trismegisti and Solitudo Criophori) and lighter areas (such as Phaethontias and Liguria). Observations of these various features with respect to the terminator led him—in error, as it later proved—to accept the idea of synchronous rotation.

Antoniadi's theory of rotation

The orbit of Mercury (M) round the Sun (S) as drawn by Antoniadi illustrates the fundamental principle on which his observations were made. Mercury's orbit is marked in days from 1 to 88, giving the requisite rotation period for the illumination, by the Sun, of only one hemisphere. With this synchronous rotation Antoniadi was able to explain the apparent fixed positions of surface markings on the disk. It was not until 30 years later that this theory was finally disproved and the actual rotation period established at 58·65 days, which means that every part of the surface is exposed to sunlight.

The "black drop" phenomenon

Transits of both Mercury and Venus occur when the planets pass across the face of the Sun and appear as black disks. Transits of Venus are rare; they occur, however, at regular intervals. Transits of Mercury are less regular but much more common; the next will be in 1986. The comparisons made by Antoniadi are still valid; thus when Venus enters transit (*left*) it is seen to be surrounded by a luminous ring; with Mercury (*right*) there is no such ring. The ring on Venus is caused by an extensive atmosphere, which is not present on Mercury. Although Antoniadi believed that Mercury had an appreciable atmosphere, he recognized that it must nevertheless be far too tenuous to produce the luminous-ring effect. When observing the transits of Venus, there was an irritating effect known as the "black drop". When Venus passed on to the Sun it appeared to draw a strip of blackness after it, and when this strip disappeared the transit was already well in progress. This was widely (and correctly) attributed to Venus's atmosphere. However, on 10 November, 1927, Antoniadi saw the same effect with Mercury (*sequence below*). This he correctly attributed to disturbances in the atmosphere of the Earth.

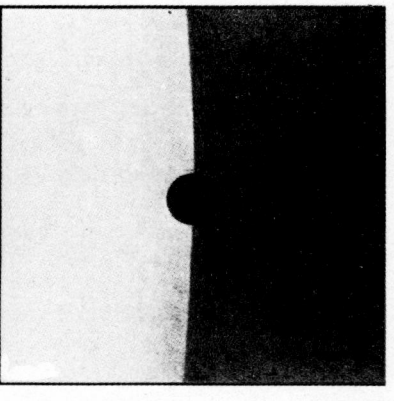

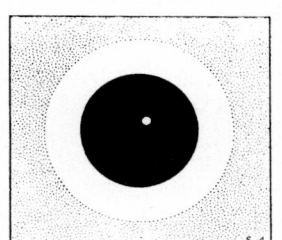

The 1868 transit

Various strange phenomena were recorded during the transits of Mercury. A greyish surrounding halo, or aureole, noted by Schröter and others, was dismissed as a contrast effect, but there were also white points on the disk of the planet as seen in this observation made during the transit of 4 November, 1868. Antoniadi rightly regarded the luminous points as being due to optical effects.

The Sun from two planets

These imaginative sketches, from *La Planète Mercure*, compare the size of the Sun as seen from Mercury (*left*) and Earth (*right*). In the former, Mercury is at its closest to the Sun (at perihelion). Although overestimating the extent of the atmosphere, in general, Antoniadi's comparison is correct: "The enormous Sun rising over the horizon of Mercury would not appear flattened, because the refraction due to the planet's atmosphere is so slight; the Sun rising over the horizon as seen from Earth is greatly flattened, because of the strong refraction caused by the Earth's atmosphere."

Atmospheric "movements"

A theory of atmospheric circulation on Mercury was a logical development of Antoniadi's belief in an appreciable atmosphere. His diagrams show (*left*) the Sun heating the atmosphere of Mercury from direction S, causing the air to rise over the sunlit hemisphere, and (*right*) the movement of warm current (continuous lines) and cool current (dotted lines) with the Sun at the zenith (Z). Although Antoniadi recognized that the density of Mercury's atmosphere was in no way comparable to Earth's, he still considerably overestimated it. In fact the density is far too low to produce such effects.

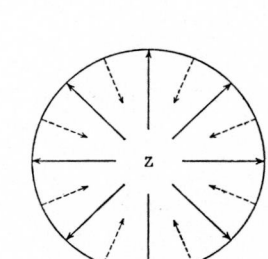

Mercury within the Solar System

Behaviour and characteristics before and after Mariner

Since the observational work of Antoniadi and his contemporaries, new techniques, especially the use of radar, have added to the basic knowledge of Mercury, which Mariner confirmed. Thus the value used by Antoniadi for the mean distance of Mercury from the Sun at 35,950,000 miles (57,850,000 km) has been revised to 35,980,000 miles (57,910,000 km). However, his eccentricity value of 0·206 for Mercury's orbit still stands; there is a considerable range between the perihelion and aphelion distances (the closest and farthest points from the Sun). The orbit is more eccentric than that of any other planet apart from Pluto, and thus has a marked effect upon surface temperatures.

A small telescope will suffice to show the phases; it is worth noting that Mercury, unlike Venus, is actually brightest when gibbous—that is, at the phase between half and full. The fact that surface markings are so difficult to observe from Earth is due partly to the fact that Mercury is never seen against a dark background, and partly to the small size of the planet. The equatorial diameter is 3,030 miles (4,880 km). This is intermediate between the diameters of the Moon (2,159 miles/3,475 km) and Mars (4,220 miles/ 6,790 km). The escape velocity is a mere 2·7 miles (4·3 km) per second, so that Mercury would not be expected to retain much in the way of atmosphere.

An iron-rich core bigger than the Moon

On the other hand Mercury is very dense. Its specific gravity is about the same as that of the Earth: 5·5 (that is to say, Mercury "weighs" 5·5 times as much as an equal volume of water would do). This suggests a heavy, iron-rich core, much larger relatively than that of the Earth. The diameter of the core has been estimated at 2,200 miles (3,500 km), making it larger than the whole globe of the Moon. One interesting result of so high a density is that the surface gravity on Mercury is virtually the same as that on Mars; that is, 0·38, where Earth is the unit of 1. The mass of Mercury is 0·055 and the volume 0·056 of that of the Earth.

Mercury has an orbital period of 87·969 Earth-days. The belief of the early observers that the axial rotation period was synchronous was not challenged until 1962. At Michigan, in the United States, W. E. Howard and his colleagues measured the long-wavelength radiations coming from Mercury, and found that the dark side was very much warmer than it should be if it received no sunlight at all.

Determining axial rotation and surface depressions

The next developments came from radar work. Basically, this involves sending out a radio pulse, "bouncing" it off a solid body and recording the "echo". When pulses are reflected from a rotating body, the "echo" is affected, and the rate of spin can be found. This was done for Mercury, initially by R. Dyce and G. Pettengill, using the huge radio telescope at Arecibo, in Puerto Rico. They found that the axial rotation period was not synchronous; the actual period is 58·65 Earth-days, or two-thirds of the orbital period. There is no area of permanent daylight or permanent night and no twilight zone.

There was also a possibility of using radar to detect elevations and depressions on Mercury. Intensive studies were carried out at the Deep Space Station at Goldstone, California, in 1972. The results were interpreted by S. Zohar and R. M. Goldstein, who described the central regions of the planet as being "dotted with hills and valleys, plus apparent craters 30 miles (50 km) across here and there". It was then thought that the hills were of a gently undulating nature. It was not until the flight of Mariner 10 that the lunar-type surface was revealed.

From time to time searches have been made for a satellite of Mercury, but it now seems unlikely that any exist. On the other hand there have been suggestions that Mercury used to be a satellite of Venus; certainly it is true that in size the ratio of Earth to Moon is not so very different from that of Venus to Mercury. However, the theory is still highly speculative.

Sizes compared
Mercury (A) is the smallest of the group of four inner planets; the other members are Venus (B), Earth (C) and Mars (D). Its equatorial diameter is 3,030 miles (4,880 km). Jupiter (E), the largest of the giant outer planets, is virtually 30 times greater in diameter; the Sun (F) more than 285 times greater. Before Mariner, measurements of Mercury's diameter, based on telescopic observation, were not exact. Original estimates varied from between 3,100 miles (5,000 km) and 2,900 miles (4,700 km). Even with the world's largest telescopes no object smaller than 500 miles (800 km) can be resolved.

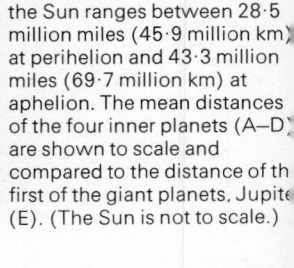

Distances compared
The distance of Mercury from the Sun ranges between 28·5 million miles (45·9 million km) at perihelion and 43·3 million miles (69·7 million km) at aphelion. The mean distances of the four inner planets (A–D) are shown to scale and compared to the distance of the first of the giant planets, Jupiter (E). (The Sun is not to scale.)

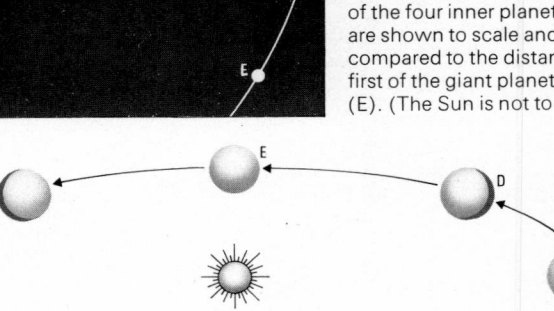

The phases of Mercury
When Mercury is between the Sun and the Earth it is new (at inferior conjunction) (A) and can only be seen in transit. As the planet moves along, a little of the daylit side begins to tilt in the Earth's direction; Mercury reaches a crescent (B), then a half (dichotomy) (C) and then the stage between half and full (gibbous) (D). At superior conjunction Mercury is full (E), but as it is then virtually behind the Sun it is barely observable. The mean synodic period is 115·9 Earth-days; that is, it takes Mercury just under 116 days to return to inferior conjunction.

Synchronous rotation disproved

Radar measurements, fully confirmed by the Mariner 10 results, show that the axial rotation period of Mercury is two-thirds of the orbital period. Mercury is shown with a fixed arrow indicating the direction of one of the two "hot poles"—that is to say, the points on Mercury's equator that directly face the Sun at perihelion. The numbers give the appropriate positions of the planet in its orbit during two revolutions round the Sun. Placed end to end, it is clear that Mercury has made three revolutions in two orbits. Each position is separated by an interval of 22 Earth-days. Both spin and orbital rotations are anticlockwise.

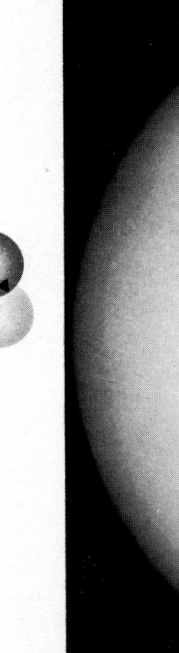

Transit characteristics

During a transit, Mercury, unlike Venus, cannot be seen with the naked eye. Telescopically it appears completely black and is much darker than any sunspots that happen to be visible. The comparison is clear in the photograph of the transit of Mercury that occurred on 7 November, 1914. During the eighteenth century it was suggested that transits of Mercury might, like those of Venus, be used to measure the length of the astronomical unit. Attempts were made to use the 1753 transit, but the results proved to be disappointing. Transits of Mercury are now regarded as being scientifically unimportant.

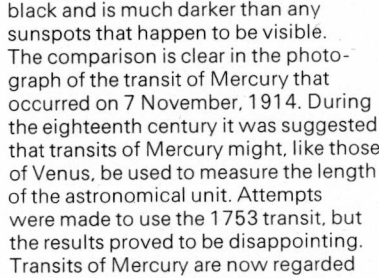

Mercury's elusiveness explained

The impression from Earth that Mercury keeps only one hemisphere turned towards us is explained by the planet's synodic period—that is, the time which elapses between successive appearances at the same phase. Because it takes Mercury two orbits to complete three revolutions, the extra revolution gives rise to a Mercurian day which is two years, or 176 Earth-days, long. This interval is approximately equal to 1½ synodic periods, so that after every three synodic periods the same face of Mercury will be seen at the same phase. Three synodic periods of Mercury add up to approximately one Earth-year,

so that the most favourable times for observing the planet recur every three synodic periods—when the same hemisphere is turned towards us. The agreement is not exact, but it was near enough to mislead all astronomers into believing in a synchronous rotation, which was not disproved until the development of radar techniques during the 1960s. The diagram shows Earth (A) and Mercury (B) over three synodic periods. Mercury's revolution is indicated by the marker. The dotted line represents the view of an observer on Earth standing at the terminator (the boundary between the day and night hemispheres of the planet).

A fleeting alignment

Because Mercury's orbit (*above*) is inclined at an angle of 7° to the ecliptic, only during relatively short periods is there any chance that the Sun, Mercury (A) and the Earth (B) will become aligned when Mercury is at its inferior conjunction. Transits can occur only during the months of May (x) and November (y). Although transits of Mercury are not common, they are more frequent than those of Venus. The tracks of Mercury across the Sun during the transits from 1960 to 2016 are shown (*left*). Clearly the transit of 1986 will last longer than the succeeding one of 1993.

Mariner 10 – the anatomy of a spacecraft

The use of gravity to alter orbits

Mariner 10's journey to the inner planets represented a pioneering achievement in space technology. Although building on techniques established in previous Mariner missions, Mariner 10 was the first vehicle to make use of the gravity of another planet—Venus—to gain its correct trajectory. It was also the first mission to transmit its television pictures as they were taken. Both the cameras and the more refined sensors incorporated on the vehicle provided a wealth of scientific information.

The design, development and construction of the whole vehicle was completed in 2½ years. The assembly is shown nearing completion, below, in good time for the launch date in November 1973. The facilities required to maintain the equipment discussed in the diagram were the propulsion system, the attitude control, temperature control, the power supply and the communications system.

Propulsion and attitude control

Because Mariner used the gravitational fields of both Venus and Mercury at each encounter to provide the main changes in direction and velocity, the propulsion unit was required to make only fine adjustments. The system provided a total velocity change of 390 ft (119 m) per second, and could be started and stopped as often as needed.

As the vehicle continued on its trajectory, it had to maintain a fixed attitude in space, with its Z axis pointing at the Sun, and its X and Y axes diagonal to the ecliptic plane. With this stable attitude, the sunshade protected the delicate instruments from the searing heat of the direct solar radiation, and the scientific instruments and the antennae were correctly pointed to make their measurements and transmit these back to Earth.

To maintain this attitude during flight required continuous control by venting minute jets of nitrogen through reaction control jets mounted at the ends of the solar panels and on the antennae and magnetometer outriggers. Two bearings ensured the correct attitude; one on the Sun, controlled by the Sun sensors, which kept the Z axis pointing at the Sun, the other on the bright star Canopus, monitored by the Canopus tracker.

Maintaining the vital transmission link to Earth

The mainstay of the temperature control was the sunshade, supplemented by multilayer thermal blankets at the top and bottom of the vehicle. Five of the eight sides were also equipped with louvred panels, which gave active temperature control by adjusting the amount of surface able to radiate into space.

The electrical power needed to operate Mariner was generated by the two solar panels carried on outriggers, which kept them clear of the shadow cast by the sunshade. A storage battery provided power for use when the solar cells were not aligned with the Sun, as happened during manoeuvres or when the Sun was occulted.

In order to function at all, Mariner had to receive commands from Earth, and transmit data back to Earth. This was achieved by the low-gain antenna and the high-gain antenna respectively; the data was received and transmitted in binary digital form.

Light in the extreme ultraviolet portion of the electromagnetic spectrum has the property of interacting strongly with most gases. Gas molecules will absorb certain ultraviolet frequencies which correspond to their characteristic wavelengths. This produces dark lines in the ultraviolet spectrum. An irradiated gas is also luminescent, emitting light at these same characteristic wavelengths. Both these effects were used to search for a Mercurian atmosphere. The occultation ultraviolet spectrometer (1) attempts to trace an atmosphere through the absorption of solar ultraviolet.

Sunlight enters the spectrometer at A and is collimated into narrow beams at B. These beams are reflected by a plane mirror (C) on which a diffraction grating is ruled. The grating splits the six beams into spectral bands (2). Detectors measure the solar ultraviolet flow at four different wavelengths centred on the absorption bands for Neon (D), Helium (E), Argon (F) and Krypton (G). Two infra-red beams (H), in the zero order reflection, time the moment of occultation.

The airglow ultraviolet spectrometer (3) detects faint light emissions from gases above the dark hemisphere. Collimation plates (A) admit sufficient light, which is reflected by a concave diffraction grating (B); this grating focuses the beam as a spectral band, which is measured by the detectors (C).

Mariner 10's radiometer measured the temperature of the surface layers of Mercury by sensing the radiation emitted. It was accurate to within ½°C. The radiometer works by measuring the minute changes of temperature in a sensitive element placed at the focal point of a telescope (A). The sensitive element is usually a thermocouple (the junction of two dissimilar metals), which generates a voltage dependent upon its temperature. For the Mariner radiometer several junctions were joined in series to form "thermopiles", which multiply the voltage (B).

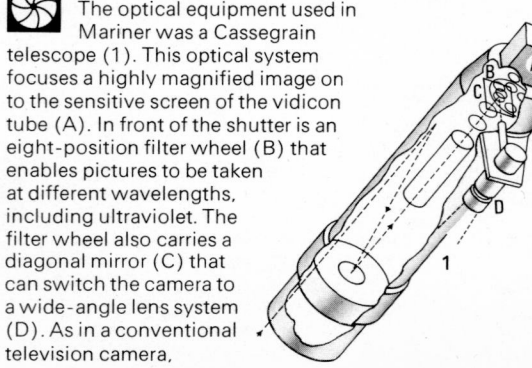

The optical equipment used in Mariner was a Cassegrain telescope (1). This optical system focuses a highly magnified image on to the sensitive screen of the vidicon tube (A). In front of the shutter is an eight-position filter wheel (B) that enables pictures to be taken at different wavelengths, including ultraviolet. The filter wheel also carries a diagonal mirror (C) that can switch the camera to a wide-angle lens system (D). As in a conventional television camera, exposure of the vidicon screen to light produces a pattern of variable charge proportional to the light intensity at each point. The pattern is scanned by an electron beam in a format of 700 lines, compared to 625 lines in typical domestic pictures. It is then transmitted to Earth in digital form as a series of binary numbers. A short burst at full power denotes 1; no transmission at all denotes 0.

Two separate telescopes and thermopiles were used, one to measure the warm regions, the other to measure the cold. Both are mounted side by side and face a diagonal mirror (C) driven by a motor (D). Through this mirror the telescopes view in succession an internal calibration surface at a known temperature, empty space and the surface of the planet in order to eliminate inconsistencies.

The magnetic fields of both Venus and Mercury were measured using the two magnetometers (1); because the magnetic environment continues between the planets, they were kept in operation during the whole flight. Each magnetometer consists of three measuring coils mounted at right angles to one another (A). Each coil measures the field strength along its own axis. By vector addition of the three components, the external field strength and direction can be reconstructed (2). Two magnetometers are used so that by comparing them both, the spacecraft's

own magnetic field is eliminated. To ensure the readings are constant, the coils are turned through 180° periodically. This operation is carried out with the flipper motor (B) for the X and Z coils, and by rolling the whole vehicle for the Y coils. In all 100 readings a second were made and transmitted back to Earth.

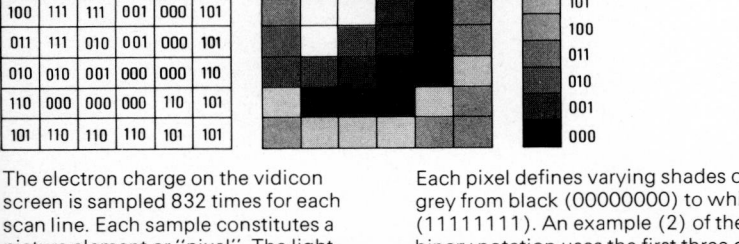

The electron charge on the vidicon screen is sampled 832 times for each scan line. Each sample constitutes a picture element or "pixel". The light intensity for each pixel is then converted into an eight-digit binary number.

Each pixel defines varying shades of grey from black (00000000) to white (11111111). An example (2) of the binary notation uses the first three digits of each pixel. Only six tones of grey are reproduced out of the possible 256 (3).

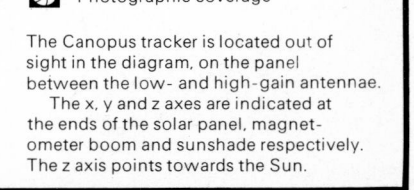

1 Airglow spectrometer
2 Canopus tracker
3 Charged particle telescope
4 High-gain antenna
5 Infra-red radiometer
6 Louvred panel
7 Low-gain antenna
8 Magnetometer (inboard)
9 Magnetometer (outboard)
10 Nitrogen reaction gas chamber
11 Occultation spectrometer
12 Plasma science
13 Propulsion unit
14 Reaction control jet
15 Solar panel
16 Sunshade
17 Television camera

Magnetic field experiment

Temperature measurement

Atmospheric detection

Photographic coverage

The Canopus tracker is located out of sight in the diagram, on the panel between the low- and high-gain antennae.
The x, y and z axes are indicated at the ends of the solar panel, magnetometer boom and sunshade respectively. The z axis points towards the Sun.

The journey of Mariner 10

A comprehensive exploration of the inner Solar System

Mariner 10's trajectory carried it more than 1,000 million miles (1,600 million km), taking it nearly three times around the Sun. From launch date until the vehicle finally lost all its nitrogen—which controlled its attitude in space —the mission lasted just 17 months. Among the scientific opportunities offered by a trajectory which made use of the gravitational field of an intervening planet, Venus, perhaps the more interesting were the comparisons drawn from data which Mariner gathered en route. Thus, using the same instruments over a short interval of time, direct comparisons were made between Earth and Moon, Venus and Mercury, throwing open new areas for speculation. At a total cost of $99,850,000—a small budget by NASA standards—our knowledge of the inner Solar System was transformed.

The pull of gravity

In escaping from the Earth's gravity (A) Mariner was slowed down enough to alter its orbit to an ellipse reaching into the orbit of Venus at closest approach to the Sun (perihelion). Without the interaction with Venus (B) it would have continued this elliptical path back out to the Earth's orbit. By approaching Venus from the outside, and swinging across in front of the planet, the gravity field of Venus was able to slow Mariner down enough for its new closed ellipse to reach farther into the orbit of Mercury (C). Once this ellipse had been established, the Sun's gravitational field automatically brought Mariner back to Mercury's orbit, although not necessarily back to the planet itself. The method used in the Venus encounter was used again at the first Mercury encounter to ensure that Mariner made more than one contact with the planet itself, not just the planet's orbit. Thus the period of Mariner's orbit round the Sun was adjusted (whilst leaving the perihelion distance unchanged) so that its returns to perihelion would synchronize with Mercury, and would thereafter continue to do so. It needed only the small velocity changes applied as trajectory correction manoeuvres (TCM) to adjust the aim between each encounter.

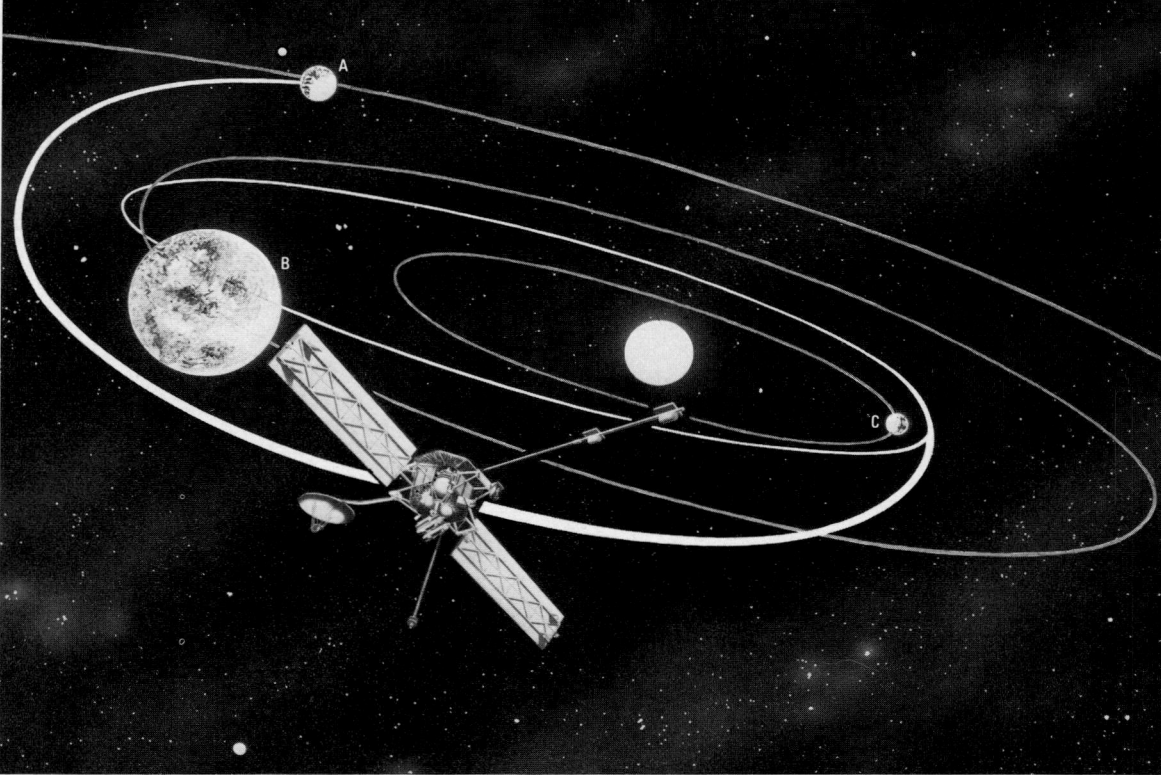

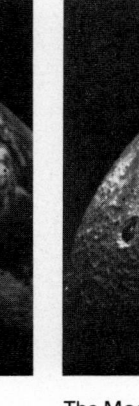

En route to the planets

The Atlas first stage accelerated Mariner 10 to 100 miles (161 km) altitude and 8,000 mph (12,900 km/h). The Centaur stage placed the craft into orbit. After 30 minutes in orbit the Centaur was reignited to increase the velocity to 25,400 mph (40,900 km/h). After escape from Earth this would leave Mariner moving more slowly than the Earth in its orbit round the Sun, by the precise amount needed to set it on course for Venus. The final Centaur burn completed, Mariner separated from the rocket and unfolded itself from its foetal launching position, illustrated in the diagram.

Experiments in space

Less than one hour after launch the Canberra Deep Space Station had control of Mariner 10, and within four hours it was correctly oriented. Obeying instructions from Canberra, Mariner rapidly came to life as its interplanetary experiments (the magnetometers, the charged particle telescopes, and later the plasma science instruments) were switched on. During the first five days of the flight five mosaics of the receding Earth were transmitted.

The Moon encounter

The launch time had been chosen to ensure that the Moon would be suitably positioned. A total of six mosaics of its surface were transmitted. These pictures gave a lead to the exposures which would be needed for taking high-resolution pictures of Venus and Mercury respectively. Detailed measurements on the Moon pictures later helped to improve the accuracy of the measurements on Mercury.

A glimpse of Venus

As Mariner approached Venus from outside its orbit, only a narrow crescent was visible. The first views were taken on 5 February only 40 minutes before closest approach. As the craft receded sunwards from Venus it could view the sunlit hemisphere, and picture-taking went on for eight days. Those taken through the ultraviolet filter showed detailed patterns of turbulent cloud markings. The rapid westward motion of these markings can be seen clearly in the three pictures, which were taken at seven-hour intervals. Although the other science experiments were designed

primarily for Mercury, they did give some useful results for Venus as well. The plasma science instrument showed that the planet's wake in the solar wind was only the same width as its ionosphere, and at close encounter the magnetometers confirmed that Venus has no detectable magnetic field. The infra-red radiometer showed that the night side of Venus is just as hot as the day side, and the ultraviolet spectrometers identified absorption lines of some constituents of the upper atmosphere. The radio occultation experiment gave information on the pressure and temperature profiles in the atmosphere.

Long journey, short encounter

The timeline scale at the bottom of the page illustrates that for 17 months in orbit Mariner spent 17 days in Mercury encounters, and only 17 hours close enough to obtain high-resolution pictures. The approach sequence of photographs was taken six days before the first close encounter. The bright spot, which was the first surface feature detected, is resolved in the last picture into a bright ray crater, subsequently named Kuiper, after the astronomer. Transmission of the full-resolution pictures every 42 seconds during the first close encounter gave a total of 647 pictures. The close dark side encounter,

not ideal for picture taking, was dictated by the need to change the orbit for further encounters. It was, however, ideal for the other experiments. Thus, the night side surface temperature was measured, both ultraviolet spectrometers were able to test for an atmosphere, and, quite unexpectedly, the magnetometers showed that Mercury had a modest magnetic field. Although Mariner left Mercury on orbit for further encounters, it was beginning to show signs of deterioration. The plasma science experiment was closed down, there was a partial failure of engineering telemetry, and during August the tape recorder failed. Following the second

encounter, which produced 300 pictures, improvisation of methods of attitude control kept Mariner alive for its third encounter. Since it was intended to discover more about the planet's magnetic field, the aim was for the closest possible dark side trajectory at high latitude, and this was attained on 16 March. Television cover was restricted because of breakdowns in the deep space network on Earth, but some very high-resolution pictures were obtained in $\frac{1}{4}$-frame format. After the third encounter the nitrogen supply for attitude control was nominally exhausted, but Mariner maintained attitude until 24 March, 1975.

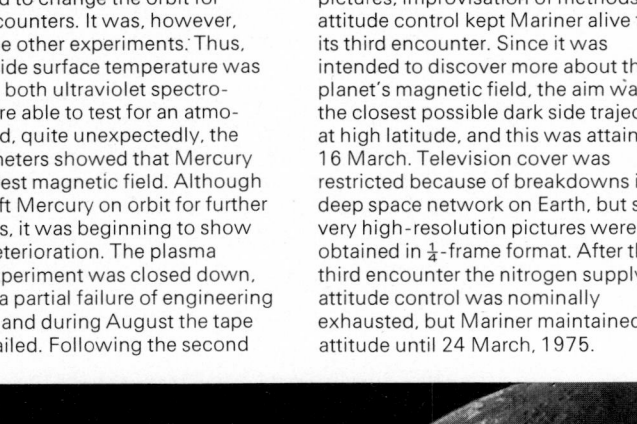

2,700,000 mi
4,350,000 km

2,190,000 mi
3,525,000 km

1,141,000 mi
1,840,000 km

590,240 mi
949,870 km

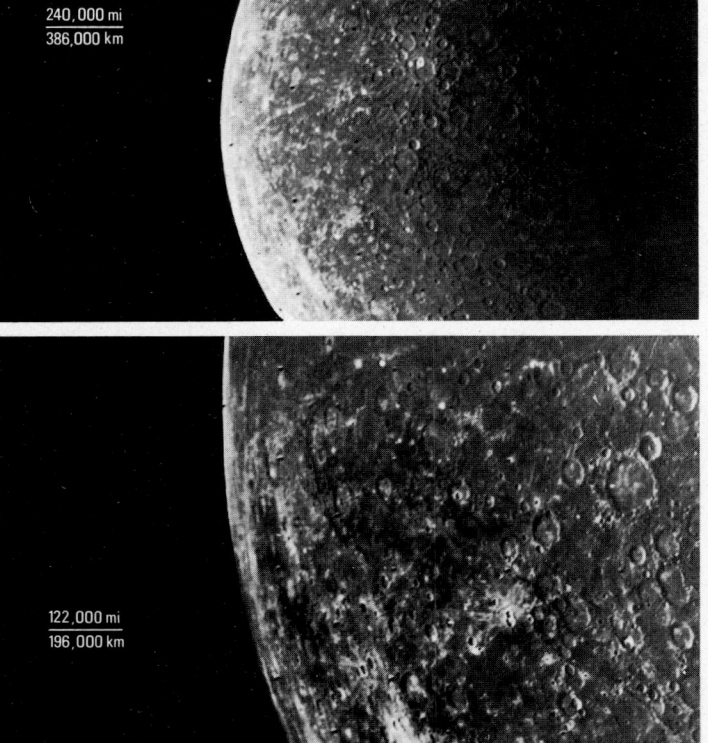

240,000 mi
386,000 km

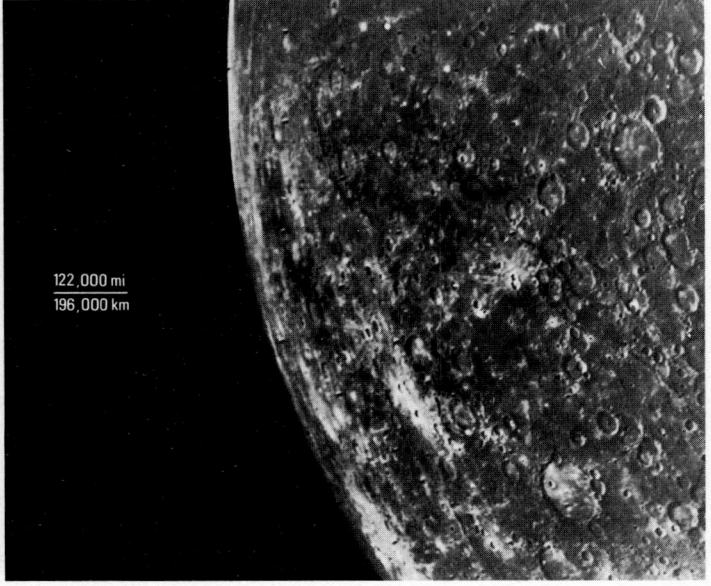

122,000 mi
196,000 km

1—7 Trajectory Correction Manoeuvres

NOV 1973 · DEC · JAN 1974 · FEB · MARCH · APRIL · MAY · JUNE · JULY · AUG · SEPT · OCT · NOV · DEC · JAN 1975 · FEB · MARCH

Earth and Moon encounter

Venus encounter

First Mercury encounter
23 March Encounter sequence starts
29 March 04.46 Computer takes control
17.13 Close encounter starts
20.42 Solar occultation starts
20.46 Time of closest approach
20.48 Earth occultation starts
20.49 Solar occultation ends
20.59 Earth occultation ends
30 March 01.01 Close encounter ends
12.46 Computer control ends
2 April Encounter sequence complete

Second Mercury encounter
17 Sept Encounter sequence starts
21 Sept 04.50 Computer takes control
17.50 Close encounter starts
20.50 Time of closest approach
23.50 Close encounter ends
22 Sept 12.50 Computer control ends
23 Sept Encounter sequence complete

Third Mercury encounter
16 March 09.12 Encounter sequence starts
20.34 Close encounter starts
23.39 Time of closest approach
17 March 00.48 Close encounter ends
22.00 Encounter sequence ends

Mapping the Mercurian surface

The equatorial stereographic projection

Mariner's three successive encounters with Mercury produced photographic coverage of only one hemisphere of the planet. This hemisphere was itself incomplete and was not ideally suited to a map using the conventional Mercator and polar stereographic projections. However, the little used equatorial stereographic projection, on which the quadrant maps are based, makes a single map of the whole illuminated hemisphere. Seen as a disk, it has the subsolar point at its centre, and the terminator (where the sun is on the horizon for an observer on Mercury) around its perimeter. Beyond the terminator the map extends into the unilluminated borders of the dark hemisphere.

Like the Mercator, the equatorial stereographic is a conformal projection. This means that shapes and directions are preserved undistorted on the map, so that circular craters are shown as circles everywhere on the surface. Since the conditions of illumination have been identical for all three encounters, the natural lighting has been reproduced on the map, with all shadows pointing directly away from the subsolar point. This shows exactly where there is no topographical data because the surface is in shadow, and also has the advantage that the maps correspond exactly in appearance to the Mariner pictures.

Construction of the maps

In normal terrestrial mapping, features are located in relation to a grid of control points fixed by a careful survey of the terrain (such as the Ordnance Survey triangulation points used in the United Kingdom). On Mercury a net of control points has been established by less direct methods.

Small, easily identifiable craters distributed over the visible surface were chosen and their precise latitudes and longitudes worked out. The first step was to measure accurately their positions on the television pictures and then to relate these image co-ordinates to the latitudes and longitudes by complex equations. These equations involved the radius of Mercury and its position and orientation in space, the position of Mariner and the orientation of its cameras in space, and the cameras' effective focal length. The values of each were recorded at the moment of exposure of each picture. Finally the equations were solved by an iterative (trial and error) method on one of the largest of the American electronic computers.

The result of all these complicated computations was a grid of 2,378 control points located to within a standard error of 6 miles (10 km) in the southern hemisphere, deteriorating to about 15 miles (25 km) at the north pole.

The Greenwich of Mercury

The 16th International Astronomical Union Commission defined the zero meridian of Mercury as the subsolar point at the first perihelion passage in 1950—that is, the point on the equator which lay directly under the Sun when Mercury was at its closest. Now that abundant topographical detail is known it is possible to relate Mercury's co-ordinate grid to its surface features. This has been done by selecting a tiny crater, Hun Kal (*below*), and assigning to it longitude 20°W. Although Hun Kal is now the Greenwich of Mercury it could not be located on the zero meridian, because this was in the dark hemisphere during the encounters.

Preparing the television cameras
The twin Cassegrain telescope vidicon cameras are seen undergoing light sensitivity calibrations in the AO Hangar Cleanroom at Kennedy Space Center. The two cameras were mounted in tandem on a scan platform, so that they could be pointed in the correct direction in space. They were used alternately: while one was photographing, the other was transmitting its picture. By equipping the cameras with powerful telephoto lenses the mission planners were able to extend the effective picture-taking period to the three hours needed for complete high-resolution cover. Extreme care in the assembly of Mariner 10 ensured its reliable operation over many months in space. The nylon coveralls worn by the technician cause static build-up, which has to be discharged to protect the vehicle's delicate instruments.

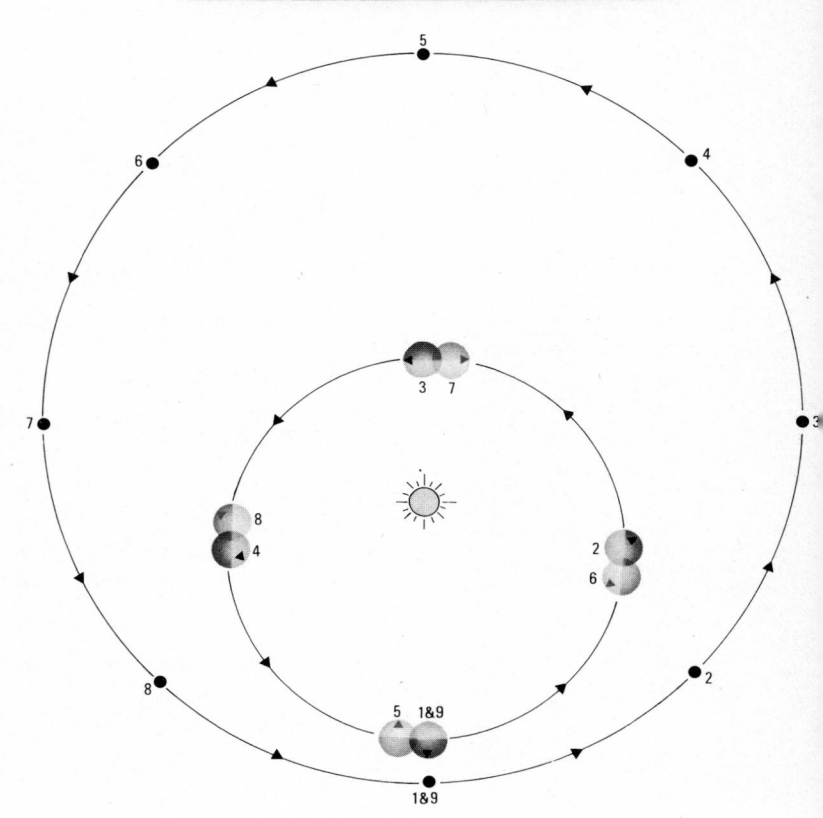

Mercury's two-year "day"
The position of both Mercury and Mariner is shown at successive 22-day intervals after the first encounter. A marker defines an arbitrary point on the planet's surface. Mariner's elliptical orbit has been adjusted so that it completes one circuit in 176 days to rendezvous with Mercury, which has meanwhile been twice round its own orbit. Mercury spins three times on its axis whilst going twice round its orbit; if it had spun only twice round on its axis during this period, Mercury would clearly always present the same face to the Sun. The extra turn on its axis causes the Sun to rise and set once, giving Mercury a day which is two of its years long. During this single day the planet has rotated three times with respect to the stars, bringing the marker back to the same orientation in position 9 that it had in position 1. Thus exactly the same hemisphere is illuminated at each encounter.

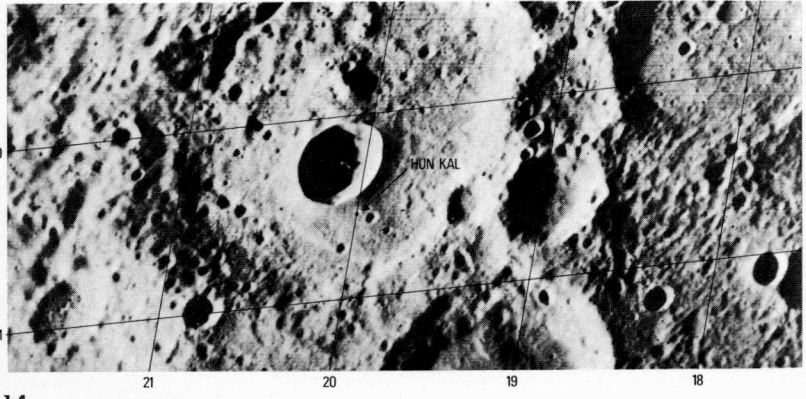

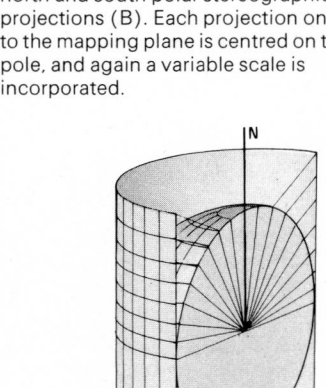
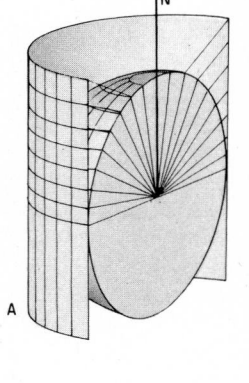

The map's missing link

Because of the geometry of the encounters not even the whole of the single hemisphere visible to Mariner could be photographed. At the first encounter Mariner caught up with Mercury approaching from below the ecliptic plane and moving inwards towards the Sun. From point A the view was a crescent (1) tilted to show the south pole (x). As it came closer the crescent narrowed rapidly until Mariner entered occultation at B. Mariner emerged from occultation and crossed the ecliptic at C. Thereafter the crescent widened rapidly to a half moon at D. From E the view was a gibbous disk (2)

tilted to show the north pole (y). Because Mariner flew by the dark side of the planet the diagonal unshaded strip never appeared above the horizon. The second encounter trajectory was parallel to the first but much closer to the Sun. At closest approach to Mercury Mariner was still well below the ecliptic plane. Thus from point F the view was a gibbous disk (3) with the south pole in the centre of the terminator (z), leaving the northern end of the strip, missed by the first encounter, still unseen. The third encounter passed along almost the same path as the first, and thus for reasons of clarity the trajectory has not been included.

Map projections: the alternatives

Mercator's projection of 1569, one of the most famous ever devised, covers a band round the equator between latitudes 70° north and south (A). To ensure that there is no distortion in shape of the surface features, the projection from the globe to the mapping plane is made using a variable scale. To map a whole planet the projection is often used in conjunction with north and south polar stereographic projections (B). Each projection on to the mapping plane is centred on the pole, and again a variable scale is incorporated.

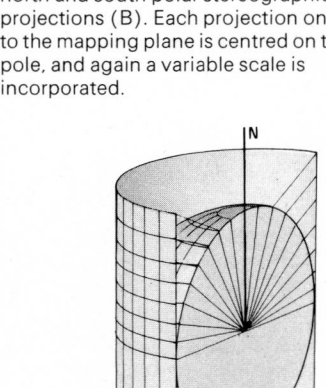

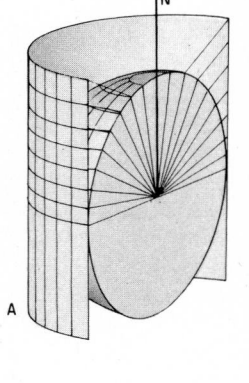

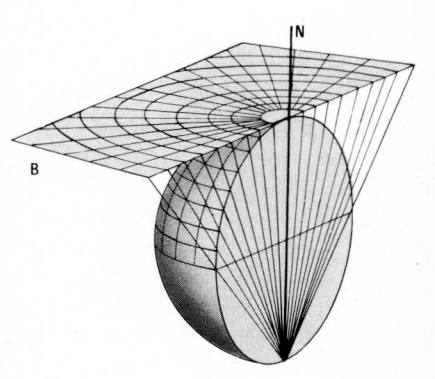

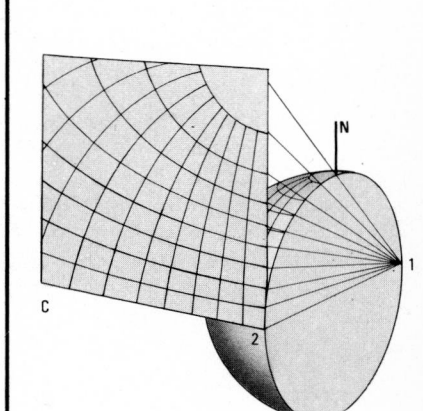

The photographic sequence

The maps were made from a total of 2650 Mariner 10 photographs, based on the first and second encounters. At each encounter the onboard computer was programmed to alter the aim of the telescope between every exposure and build up a series of picture mosaics each giving complete coverage of the surface. However, immediately before and after closest approach there was not enough time for this; only partial mosaics were obtained in which scattered small areas are shown close up, at very high resolution. The examples show complete (A) and partial (B) mosaics in the approach sequence of the

first encounter, and partial (C) and complete (D) mosaics as Mariner receded from the planet. Because Mariner passed so close to Mercury at this encounter it saw one hemisphere when approaching and the opposite one during retreat. The viewing angle changed very rapidly from the one to the other as the vehicle swung round the night hemisphere. The second encounter passed by at a much greater distance. There was no period of partial mosaics, and during the whole near-encounter sequence the viewpoint was changing steadily. The series of mosaics obtained covered the centre of the disk, and supplemented the first encounter results.

When used to map exactly one illuminated hemisphere, a stereographic projection centred on the equator instead of the pole has special advantages (C). Again, like the Mercator projection, the scale is variable, being doubled at the terminator, which helps to show the detail brought out by the low Sun angle in these regions. The equatorial stereographic projection used for this atlas is formed by projection from Mercury's antisolar point (1) at the time of the encounters (latitude 0°, longitude 280°) on to a mapping plane tangential to the subsolar point (2).

The charted hemisphere of Mercury

The Moon and Mercury compared

A glance at the map of the complete hemisphere of Mercury shows that craters dominate the whole scene. Craters exist on all the inner planets of the solar system: Mercury, Venus, the Earth (to a minor extent) and the Moon, Mars and the dwarf satellites Phobos and Deimos. Yet the structures are not identical; Venus, with its remarkably dense atmosphere, seems to have craters that are large and shallow; on Mars the scene is affected by the presence of giant volcanoes far surpassing anything known elsewhere.

On the other hand there is a very obvious similarity between the craters of Mercury and those of the Moon. For this reason, direct comparisons between the surface details on each have been made on the following pages of photographs. As well as the type of crater, the general laws of their distribution on the Moon also apply to Mercury; for example, when one crater breaks into another it is usually the smaller which is the intruder.

The most obvious difference between the lunar and Mercurian scenes is that Mercury lacks the very well-defined dark plains which, on the Moon, are known as seas (*maria*). It has been said that the Mercurian plains (*planitia*) are of the same general type, but it is undeniable that they are visibly much less obtrusive. Equally, the only major mountains (*montes*) discovered by Mariner 10 are those that form the north and south rims of the largest feature on Mercury, the Caloris Basin (Planitia Caloris) in the north-west quadrant. However, like the comparisons made between the near and far side of the Moon, in all probability the unknown parts of Mercury are essentially similar to those charted here.

The official nomenclature

With the first results from Mariner 10 it was clear that the old nomenclature of Antoniadi had to be abandoned. The whole matter was made the responsibility of the International Astronomical Union (IAU), the controlling body of world astronomy. Six major classes of topographical features needed to be named: craters, scarps (*dorsa*), mountains (*montes*), plains or basins (*planitia*), ridges (*rupes*) and valleys (*valles*).

The most significant of the plains, the Caloris Basin, literally, the "hot basin", is so called because of its location near one of the two hot poles. The other plains take their names from the word for Mercury in various languages; thus Suisei Planitia is the Japanese form. The exception is Borealis Planitia; this name originates from its proximity to the north pole.

In recognition of the work by Earth-based radar, the valleys are named after radar installations; for example, Arecibo Vallis is taken from the huge radio telescope at Arecibo, Puerto Rico. The scarps have been named after famous ships of discovery and exploration, such as Endeavour Rupes and Santa Maria Rupes, which recall the association of Mercury with travel and commerce in Greco-Roman mythology. The names selected for the ridges commemorate famous astronomers who have been particularly associated with Mercury: notable examples are Schiaparelli and Antoniadi.

Craters are in general named after men and women who have made great contributions to the arts and humanities. Writers include Asvagosha, Homer and Li Po; artists Chao Meng-Fu, Sinan and Titian; and composers Bach, Mozart and Ts'ai Wen-Chi. One exception is Kuiper; the crater is named in honour of Gerard P. Kuiper, who was so prominent in planetary probe research. The other exception is the name Hun Kal, allotted to the very small crater accepted as the Greenwich of Mercury. Hun Kal is the Mayan word for the figure twenty; by definition, the 20th meridian of longitude on Mercury passes through the centre of this crater.

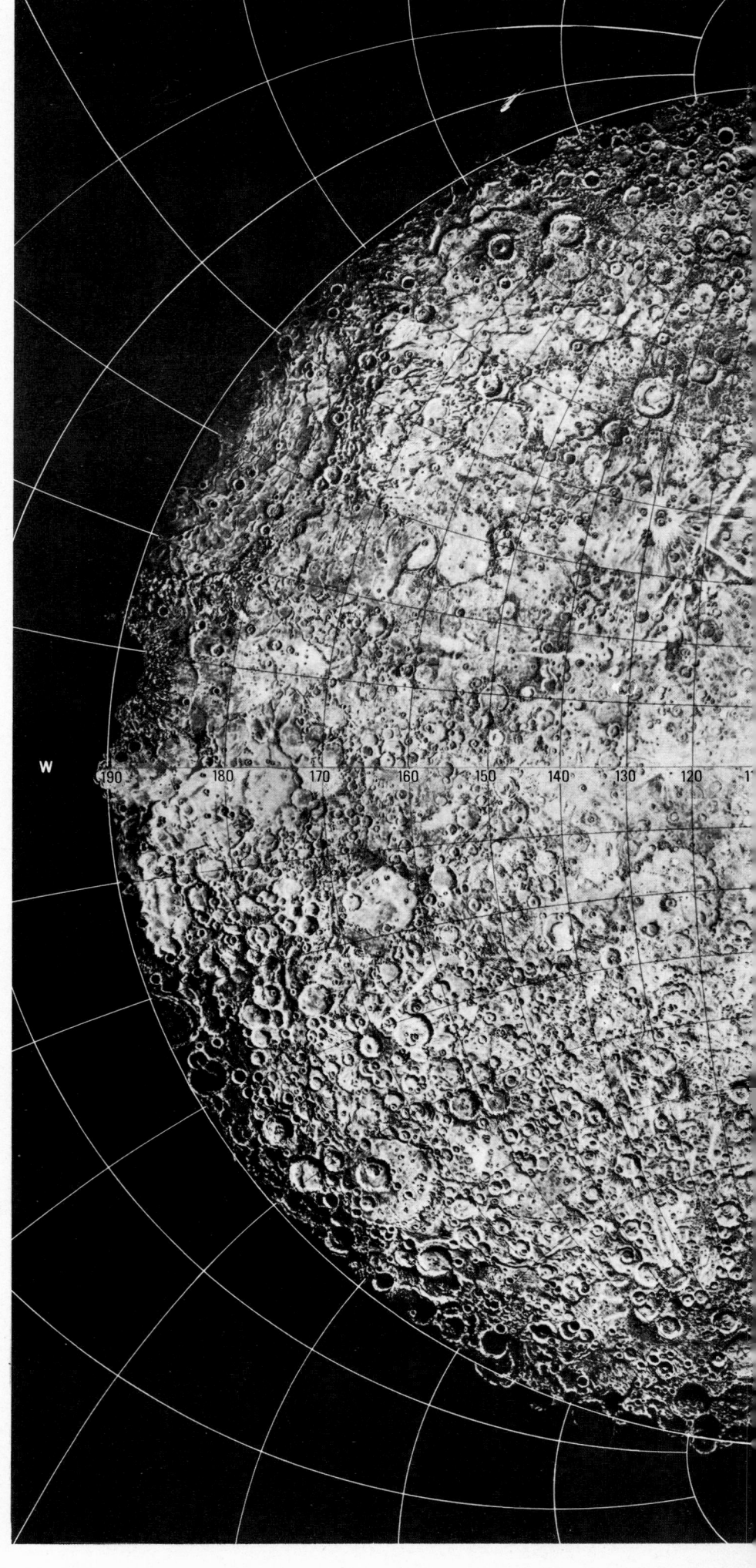

The subsolar point

The complete view of this hemisphere of Mercury clearly distinguishes the subsolar point from surrounding regions. Defined as the point on the equator that lies directly under the Sun, the subsolar point—located at the centre of the hemisphere at latitude 0°, longitude 100°—is under vertical solar illumination. Thus the apparent paucity of craters in this area is caused by the absence of any shadows to define their walls.

Location of the surface pictures

Each of the four quadrant maps is accompanied by four pages of surface pictures. The pictures are numbered consecutively and these numbers are reproduced in each of the quadrant gazetteers alongside the feature or features appearing in the picture. The captions to each picture list those features that have so far been officially named by the IAU. Where names have yet to be assigned the latitude and longitude of the centre of the picture concerned is given. The spelling used keeps strictly to the IAU nomenclature, so that less common spellings of proper names, such as Chaikovskij and Tolstoj, occur. All crater names are labelled from A to Z on each picture; all other features are spelt out.

Picture information

Every picture is oriented with north at the top, and a bar scale in miles and kilometres is included when possible; where a perspective picture has been created by the acute viewing angle of the Mariner telescopes, the scale is necessarily omitted. Close-ups of certain pictures are indicated by an outline of the appropriate area. Where two close-ups are taken from the one picture, they are distinguished by a solid and dotted box outline. In some cases the boundaries of irregular features that are ill-defined have also been picked out with a dotted line.

Conversion factors

For mile and kilometre conversions in the main text, approximations have been made where the original figure is not exact; thus the accuracy for each figure is to the same decimal place. However, in each quadrant gazetteer, because the official dimensions have been given in kilometres, the conversion to miles is exact.

Using the quadrant scale

The scale below (for the map of the complete hemisphere) is similar to those which accompany each quadrant page. With a centimetre ruler, measure the distance from the subsolar point to the centre of the feature concerned. Locate this measurement on the horizontal centimetre scale. Measure the feature concerned. Place the ruler vertically at the point previously located on the centimetre scale. On the ruler read up from the base line the number of centimetres that the feature measures against the kilometre scale.

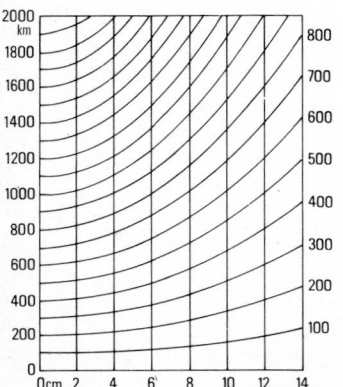

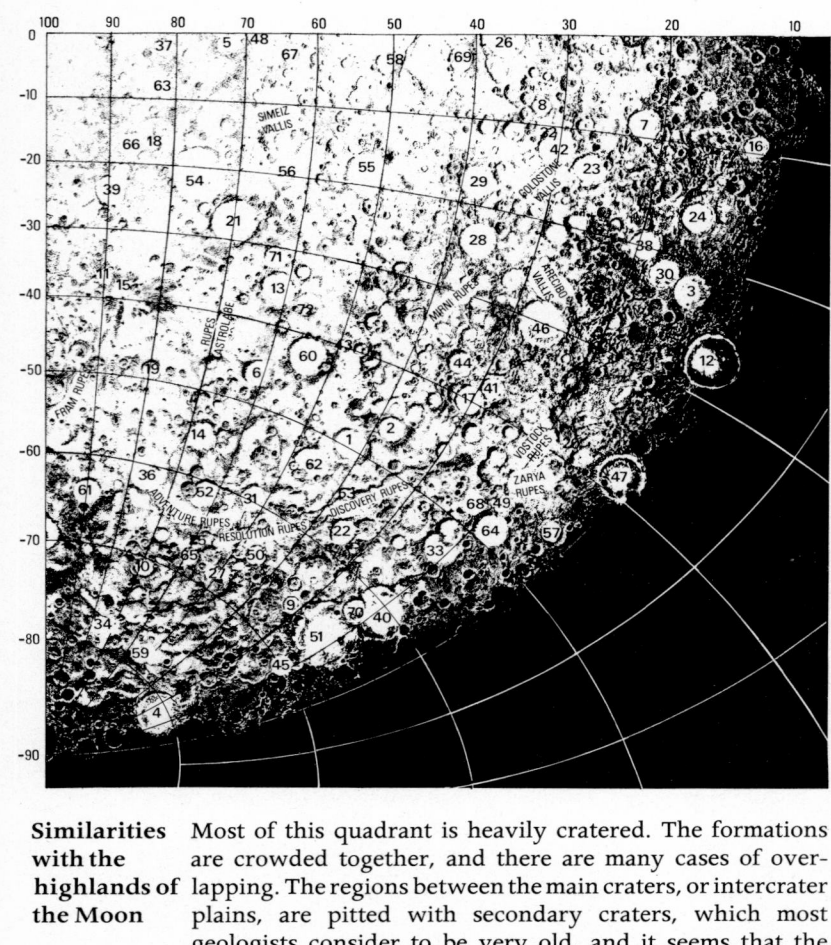

Craters	Latitude	Longitude	Diameter (miles)	Diameter (km)	Caption No.
1 Africanus Horton	−50·5	42	75	120	
2 Andal	−47	35·5	56	90	
3 Balagtas	−22	14	62	100	14
4 Boccaccio	−80·5	30	84	135	
5 Boethius	−0·5	74	81	130	
6 Bramante	−46	62	81	130	
7 Brunelleschi	−8·5	22·5	87	140	4
8 Byron	−8	33	62	100	
9 Callicrates	−65	32	40	65	
10 Camões	−70·5	70	43	70	
11 Carducci	−36	90	47	75	
12 Cario	−27·5	10	99	160	14
13 Chekhov	−35·5	61·5	112	180	
14 Coleridge	−54·5	66·5	68	110	
15 Copley	−38·5	86·5	19	30	8
16 Dvořák	−9·5	12·5	50	80	4
17 Equiano	−39	31	50	80	
18 Futabatei	−16·5	83·5	34	55	
19 Ghiberti	−48	80	62	100	8
20 Guido d'Arezzo	−38	19	31	50	12, 13
21 Haydn	−27·5	70	143	230	
22 Hesiod	−58	35·5	56	90	
23 Hiroshige	−13	27	87	140	9
24 Hitomaro	−16	16	65	105	4, 5
25 Holberg	−66·5	60	41	66	
26 Homer	−1	36·5	199	320	
27 Horace	−68·5	50	30	48	
28 Ibsen	−24	36	99	160	1
29 Imhotep	−17·5	38·5	99	160	1
30 Kenkō	−21	16·5	56	90	14, 15
31 Khansa	−58·5	52	62	100	
32 Kuiper	−11	32	25	40	9
33 Kurosawa	−52	23	112	180	
34 Li Ch'ing Chao	−77	75	37	60	
35 Lu Hsun	0·5	23·5	59	95	
36 Ma Chih-Yuan	−61	77	106	170	
37 Machaut	−1·5	83	65	105	
38 Mahler	−19	19	62	100	4, 14
39 Matisse	−23·5	90	130	210	
40 Mendes Pinto	−61	19	106	170	
41 Mofolo	−38	29	56	90	
42 Murasaki	−12	31	78	125	9
43 Nampeyo	−39·5	50·5	25	40	
44 Neumann	−37·5	34	62	100	
45 Ovid	−69·5	22	25	40	
46 Petrarch	−30	26·5	99	160	1, 3
47 Pigalle	−38	11·5	81	130	12
48 Polygnotus	0	68·5	81	130	
49 Po-ya	−45·5	19	56	90	6, 7
50 Puccini	−64·5	46	68	110	
51 Pushkin	−65·5	23	124	200	
52 Rabelais	−59·5	62·5	81	130	
53 Rameau	−54	38	31	50	11
54 Raphael	−20·5	76·5	217	350	
55 Renoir	−18	52	137	220	
56 Repin	−19	63	59	95	10
57 Rilke	−45·5	12·5	43	70	
58 Rudakī	−3·5	51·5	74	120	
59 Sadī	−78·5	55	37	60	
60 Schubert	−42	54·5	99	160	
61 Sei	−63·5	89·5	81	130	
62 Shevchenko	−53	47	81	130	
63 Snorri	−8·5	83·5	12	20	
64 Sōtatsu	−48	18·5	81	130	6
65 Spitteler	−68	59	41	66	
66 Sullivan	−17	87	84	135	
67 Thākur	−3·5	64	71	115	
68 Tintoretto	−47·5	23	37	60	6
69 Titian	−3	42·5	71	115	
70 Tsurayuki	−62	22·5	50	80	
71 Unkei	−31	62·5	68	110	
72 Wergeland	−37	56·5	22	35	

Rupes					
Adventure	−64	63			
Astrolabe	−42	71			
Discovery	−54	38			11
Fram	−57	94			
Mirni	−37	40			
Resolution	−62	52			
Vostock	−38	19			12
Zarya	−42	22			

Vallis					
Arecibo	−27	29			1, 2
Goldstone	−15	32			9
Simeiz	−13	66			10

Similarities with the highlands of the Moon

Most of this quadrant is heavily cratered. The formations are crowded together, and there are many cases of overlapping. The regions between the main craters, or intercrater plains, are pitted with secondary craters, which most geologists consider to be very old, and it seems that the quadrant has less than its fair share of younger craters and basins. There are, however, three ray-craters which are presumably among the most recently formed of the features: these are Copley (15), Kuiper (32) and Snorri (63). Kuiper is of special interest. It was actually the first feature to be identified on Mercury during the first approach of Mariner 10, and it is very prominent. It interrupts the wall of a larger crater, Murasaki (42).

The obvious resemblance between the intercrater plains and the highlands of the Moon implies that the same process of differentiation was at work—that is to say that during the formation of both planetary bodies, the condensed material separated into a heavy core and a lighter crust. However, this process seems to have been more marked on Mercury than on the Moon.

There is a different kind of terrain in the area between latitudes 20° and 40°S and longitudes 10° and 40°W, which is more hilly and lineated. It is interesting to note that this area is antipodal to the Caloris Basin (Planitia Caloris), though it is smaller. Since this hilly and lineated terrain is found only in this region, it has been suggested that it may have been produced by seismic effects being focused through the globe of Mercury at the time of the formation of the Caloris Basin; however, this is still very speculative.

Obliteration of peaks and inner rings

Everywhere in the quadrant the floors of many of the craters and basins have been flooded with smooth plains material, so that any central mountain peaks and inner rings have been obliterated. A typical example is the large crater Pushkin (51) and its companion Mendes Pinto (40); others include Raphael (54); Haydn (21); Pigalle (47)—a crater situated too near the terminator to be clearly shown—and Petrarch (46). Eight of the 16 named scarps are in this quadrant. They are thought to have been formed by crustal compression, and provide confirmation of the ancient character of the terrain. The quadrant also contains three of the four named valleys; Goldstone and Arecibo are secondary crater-chains, but Simeiz seems to be a prominent terrace on the wall of a large unnamed crater.

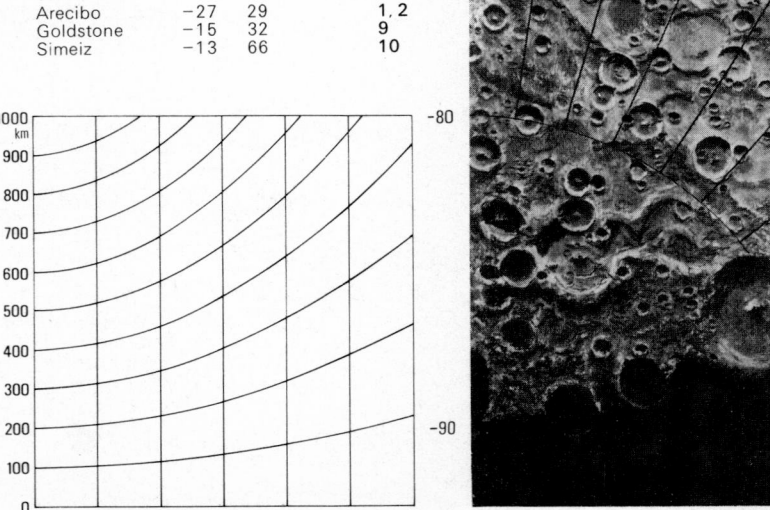

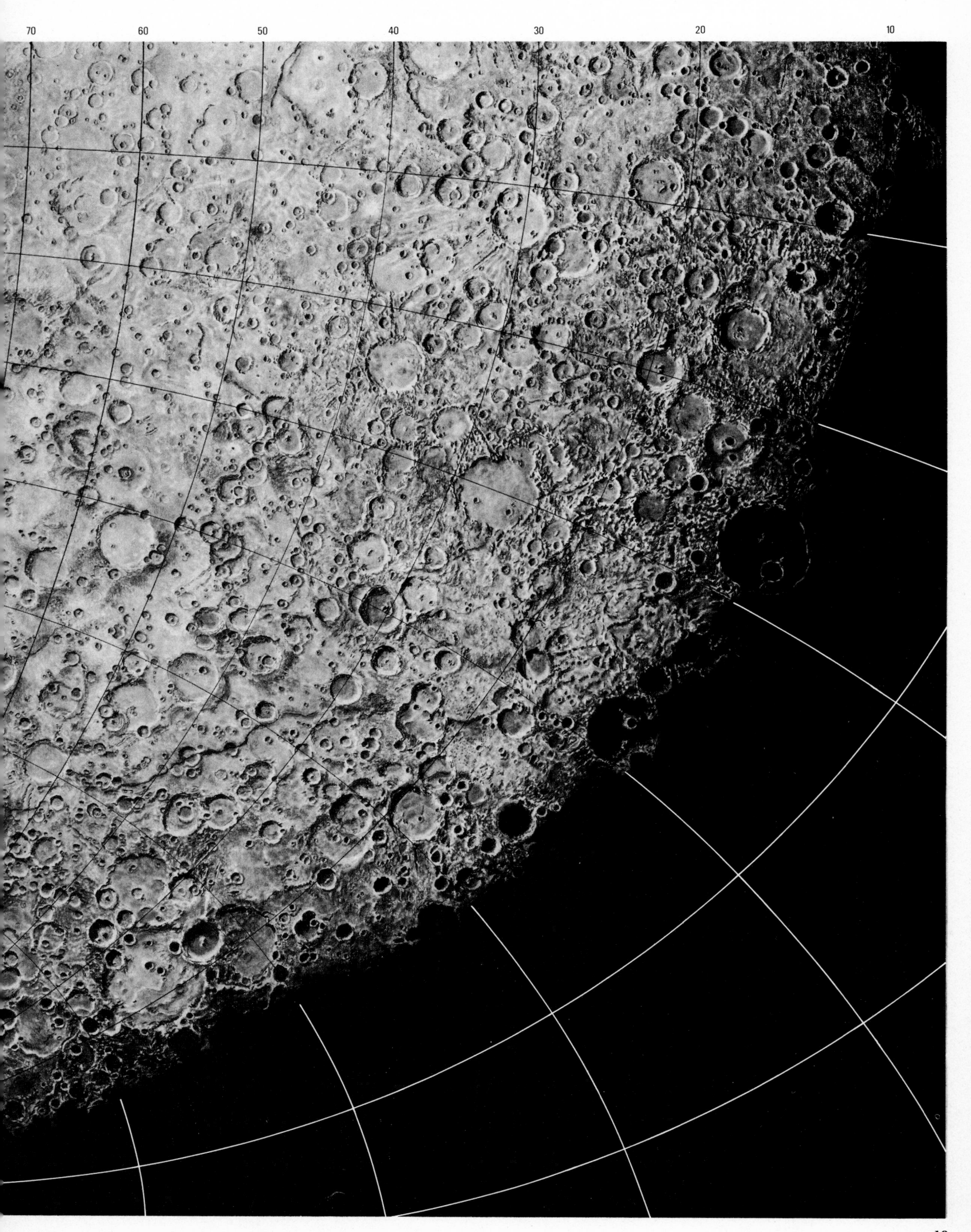

South-east panorama 1

1

2

3

4

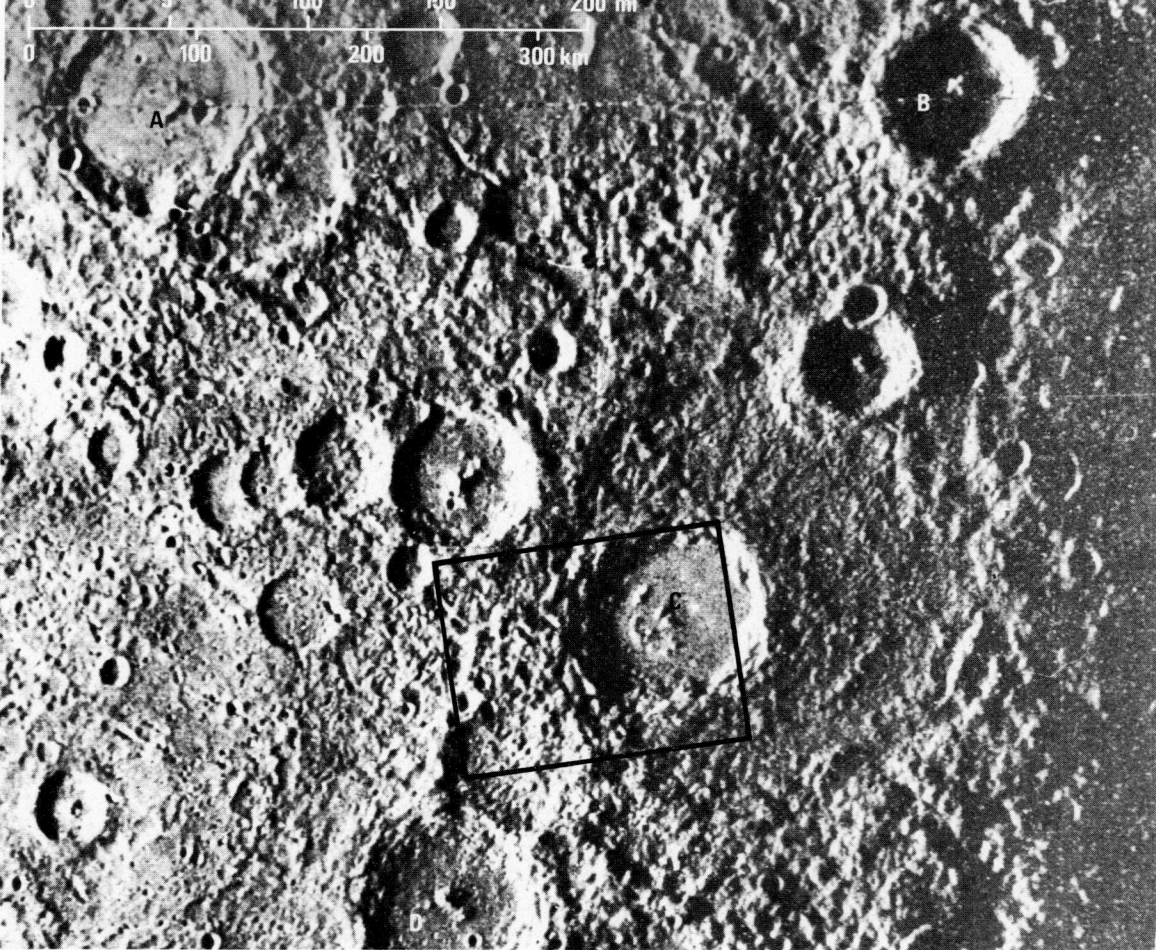

1 Ibsen (A), Imhotep (B), Petrarch (C), Arecibo Vallis. The three craters are comparal in size. Ibsen and Imhotep have smooth, flat floors and seem to have been flooded wit lava; the crater-chain linking the two is an interesting feature. Petrarch is similar, and is joined to the 55-mile (90-km) unnamed crater to its north-west by the Arecibo Vallis.

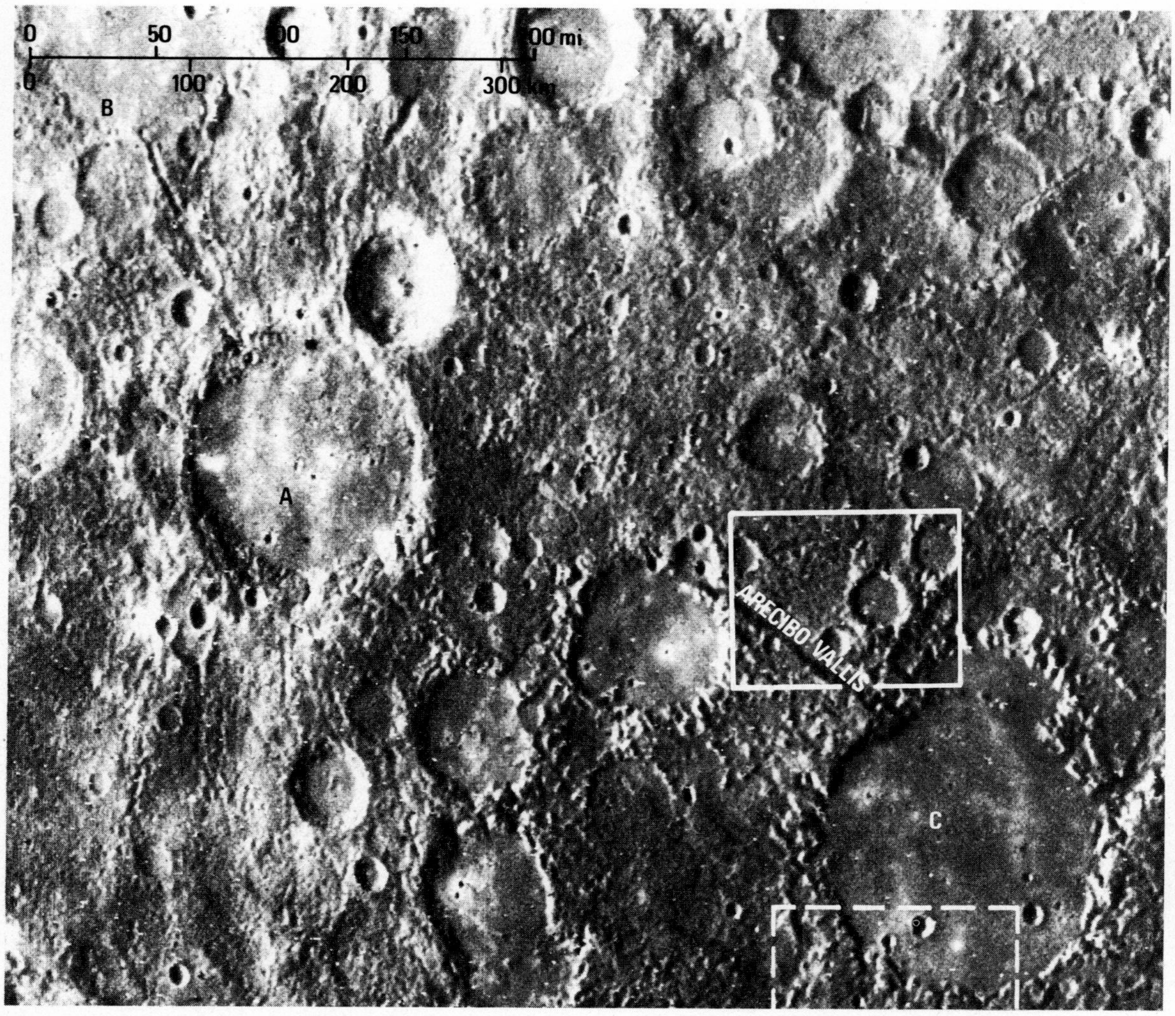

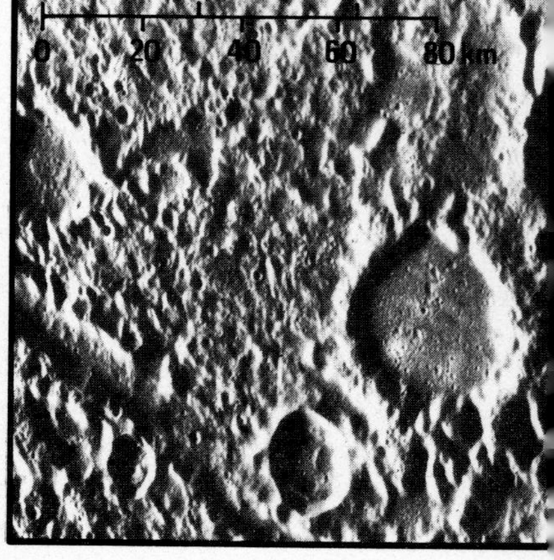

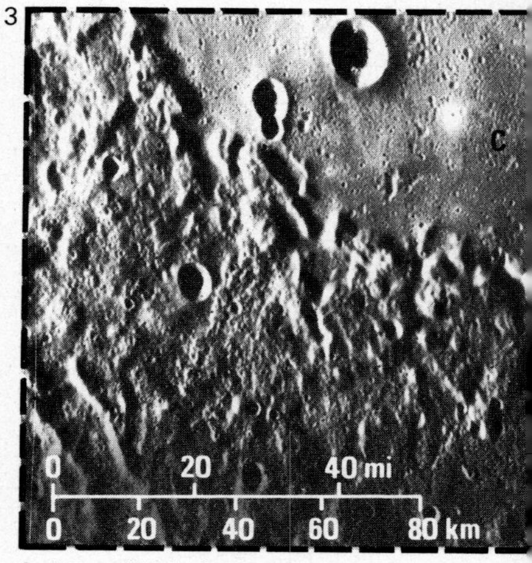

2 Arecibo Vallis. The sides of the valley are noticeably steep, and the floor is covered with the same material as the interior of Petrar

3 Petrarch (C). This detailed view shows tha the floor is pitted with smaller craters similar to those of the lunar plains.

4 Brunelleschi (A), Dvořák (B), Hitomaro (C Mahler (D). Brunelleschi has a fragmentary mountain ring, with its south-east portion mo clearly marked; Dvořák and Mahler have well-defined central mountain structures. The unnamed crater between Hitomaro and Mahl is very similar to many of the lunar craters.

5 Hitomaro (C). Hitomaro, with its off-centre irregular cluster of mountain peaks, is itself contained in a larger, more ancient structure.

5

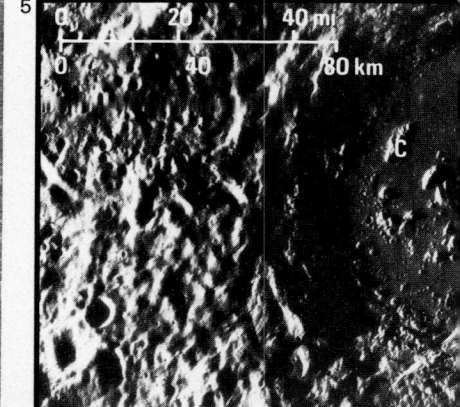

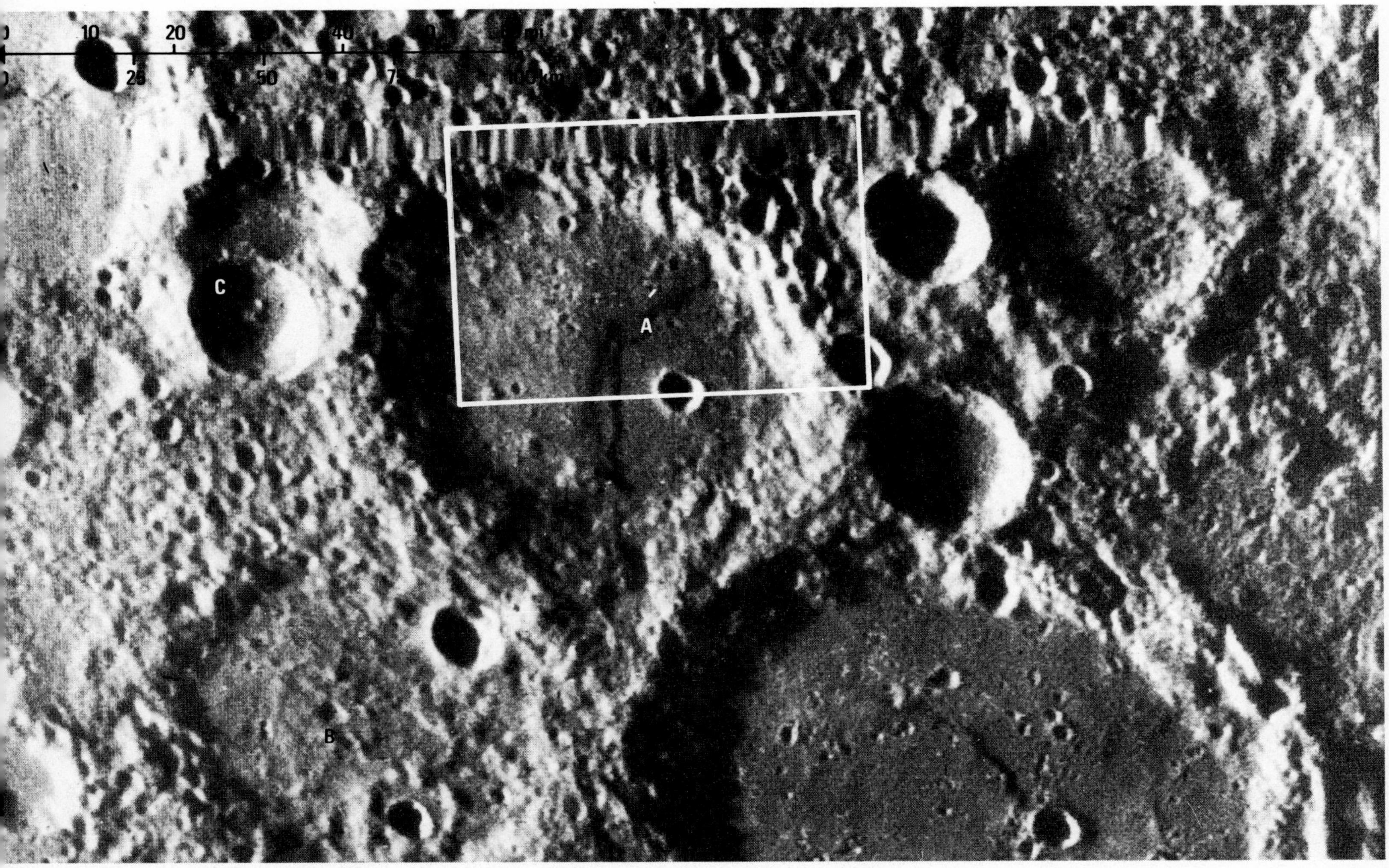

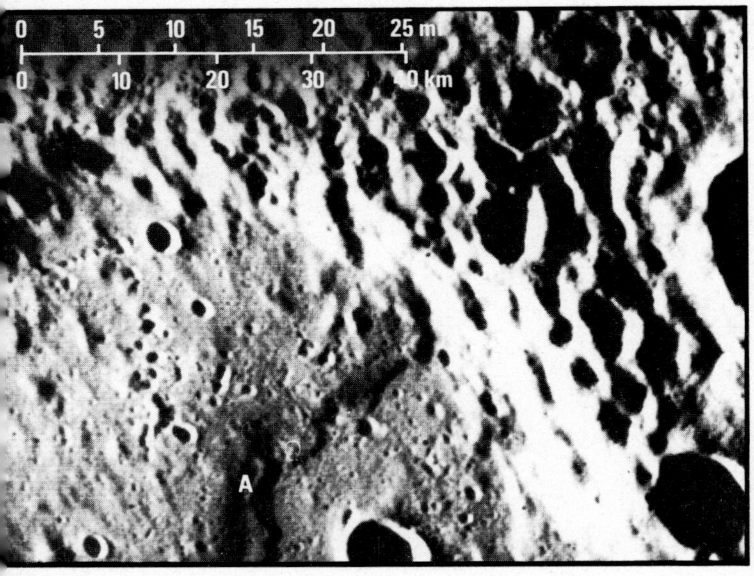

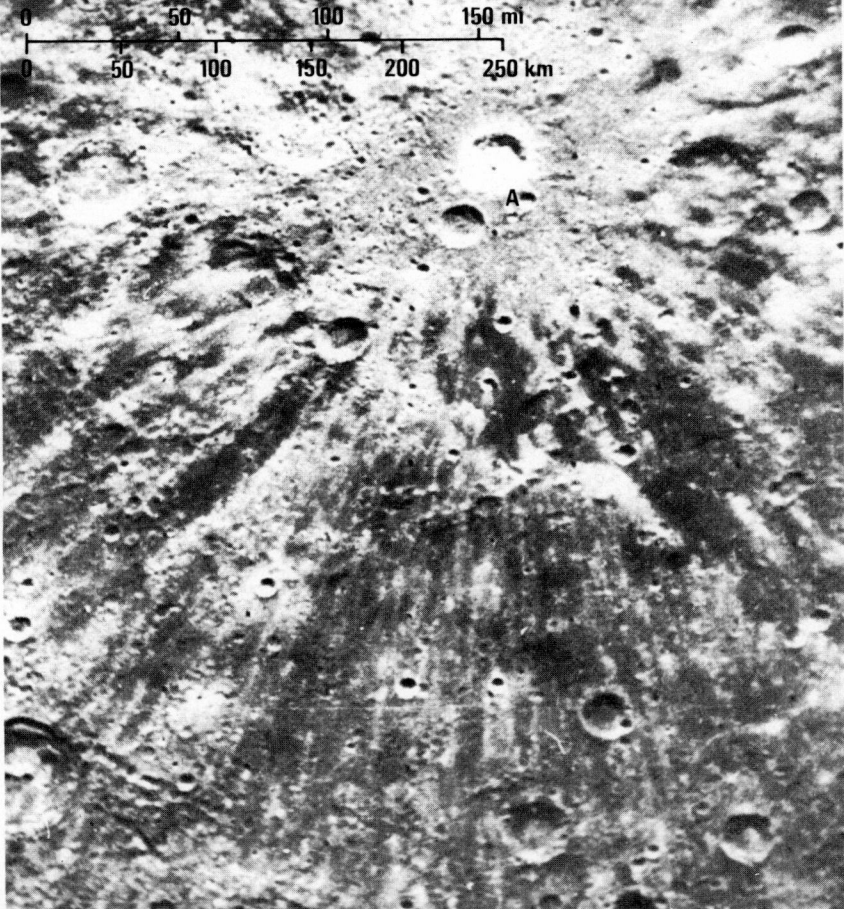

6 Po-ya (A), Sōtatsu (B), Tintoretto (C). The most imposing formation in this picture is Sōtatsu, which is not much smaller than the famous lunar walled plain Ptolemaeus. It has no central mountain group, though there are some craterlets and hills; the walls are noticeably terraced, and the rim is irregular. Adjoining it, and clearly associated with it, is the smaller but basically similar crater Po-ya. The crater-pair outside the wall of Po-ya is interesting; the smaller member, with its central peak, overlaps a larger but shallower formation. Tintoretto has relatively low walls and a rough floor.

7 Po-ya (A). The scarp inside Po-ya, restricted to the crater floor, seems to be a lava-flow front rather than a compression scarp.

8 Copley (A), Ghiberti (B). Copley is a prominent ray-crater. It is very bright, and in size rather smaller than the brilliant lunar crater Aristarchus; close beside it is a crater of similar size, which has a central peak but is not itself a ray-centre. The light-coloured rays from Copley extend for some 250 miles (400 km) across the dark plains to the south-east, reaching the northern ramparts of the much larger formation Ghiberti.

South-east panorama 2

9

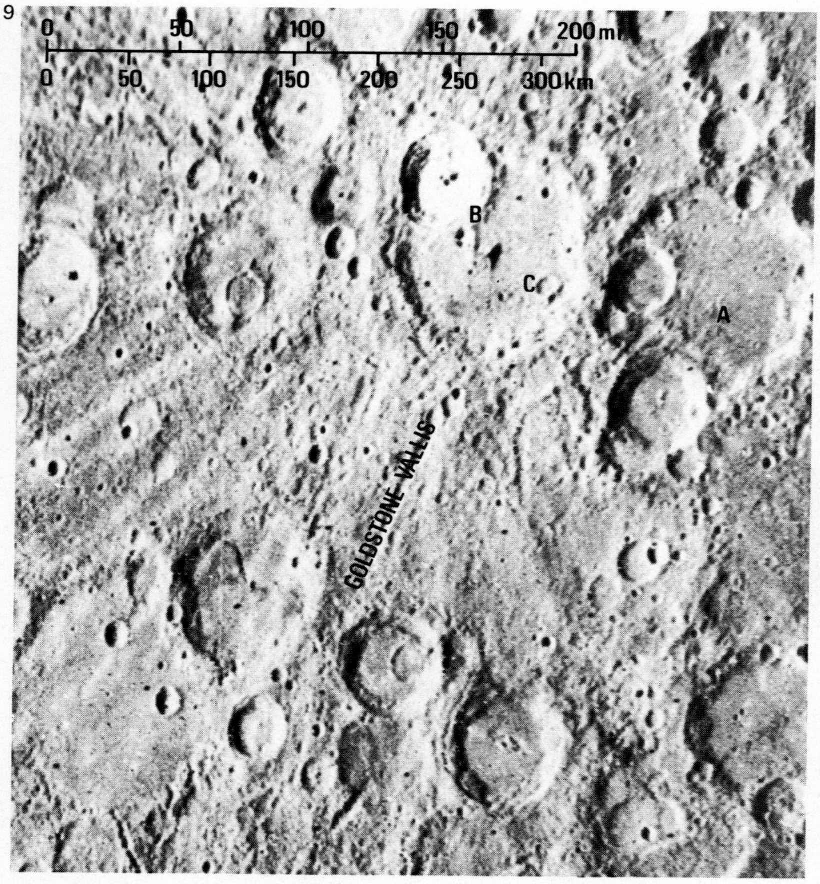

10

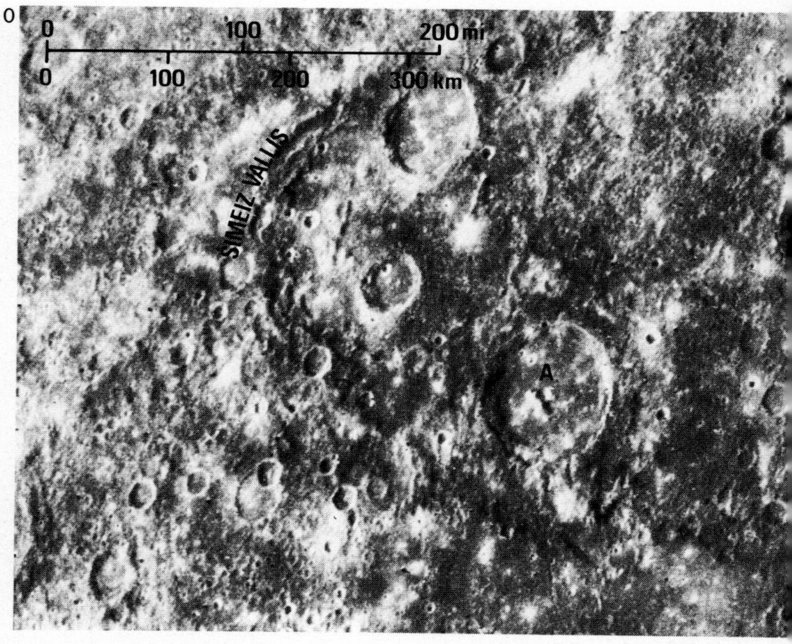

9 Hiroshige (A), Kuiper (B), Murasaki (C), Goldstone Vallis. The most prominent crater in this picture, though not the largest, is Kuiper, which is a ray-centre. It interrupts the wall of the larger Murasaki, which has a central peak; adjoining Murasaki is Hiroshige, which has no central elevation. The 75-mile (120-km) Goldstone Vallis is really a crater-chain.

10 Repin (A), Simeiz Vallis. Repin is regular, with a central peak. The Simeiz Vallis is merely a terrace in the wall of a large, obscure formation.

11 Rameau (A), Discovery Rupes. The magnificent Discovery scarp cuts through the larger crater, Rameau; it seems to have been formed by uplift due to crustal compression.

11

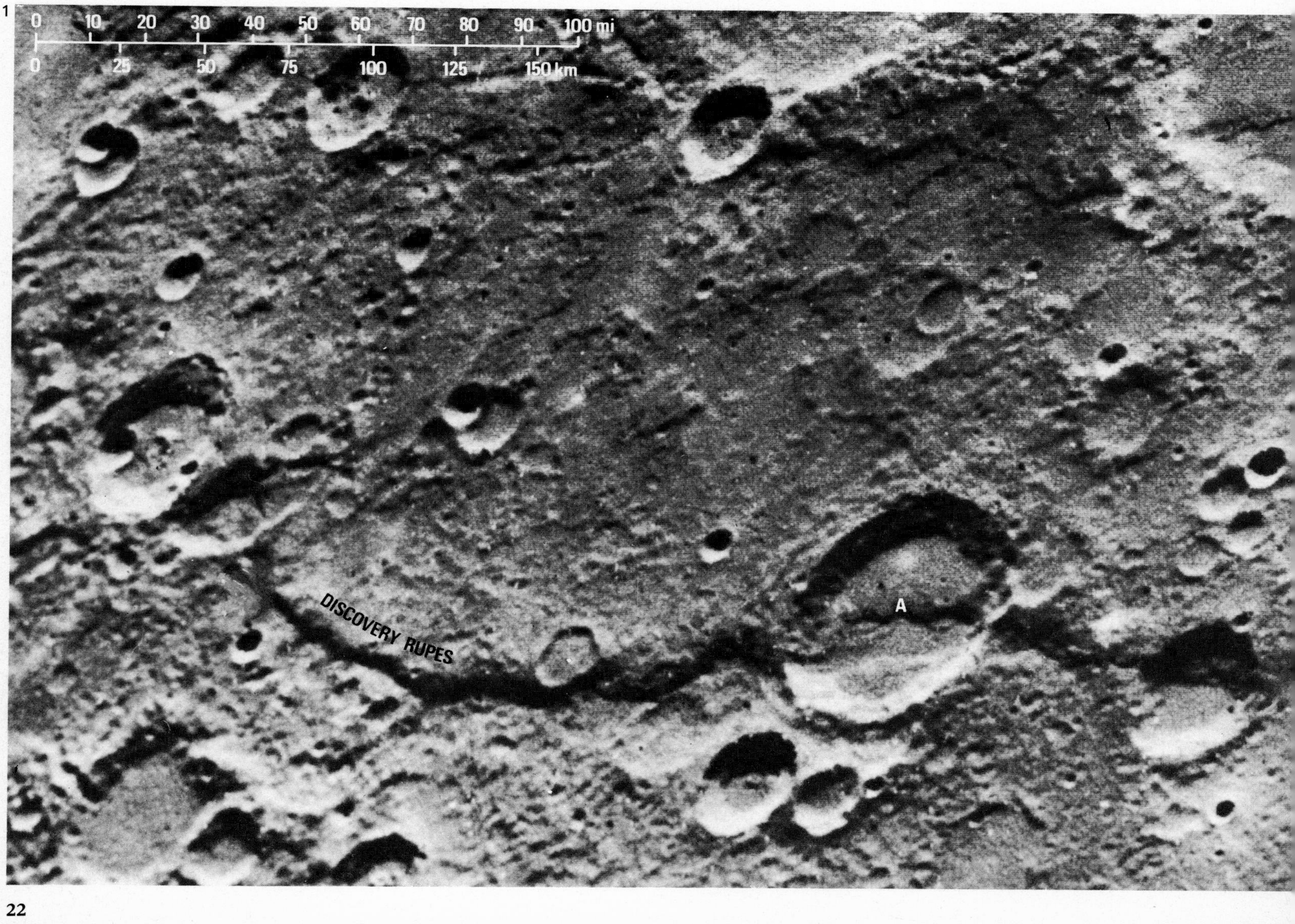

12 Guido d'Arezzo (A), Pigalle (B), Vostock Rupes. The large crater Pigalle has no central peak, and in this picture there is considerable shadow on the floor; the interior contains various small craterlets, including an interesting line north of the centre of the formation. Guido d'Arezzo is crossed by a scarp that is part of the Vostock Rupes.

13 Guido d'Arezzo (A). This close-up view shows a terraced face to the scarp. The scarp has obviously distorted the outline of the crater, implying that the former is a compression fault.

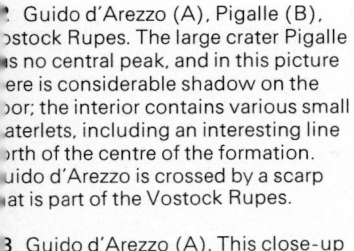

14 Balagtas (A), Cario (B), Kenkō (C), Mahler (D). When this picture was taken, the large formation Cario was on the terminator, and is therefore filled with shadow. A central peak is visible, and the floor contains considerable detail, including a large crater. The other three named craters, Balagtas, Kenkō and Mahler are of similar size and are aligned; Kenkō has been slightly distorted by Balagtas, and the arrangement is reminiscent of Walter, Regiomontanus and Purbach on the Moon. Both Balagtas and Mahler have central elevations of some complexity, and their walls are partly terraced. The secondary crater-chains crossing Kenkō from Mahler can also be traced inside Balagtas. Below these three aligned formations is a pair of craters, as yet unnamed; the larger has a central peak, while its companion is deeper, with indications of wall-terracing.

15 Kenkō (C). This close-up picture shows the secondary crater-chains which come from Mahler and extend across the floor of Kenkō. The alignments are obvious, and many of the craterlets overlap.

The south-west quadrant

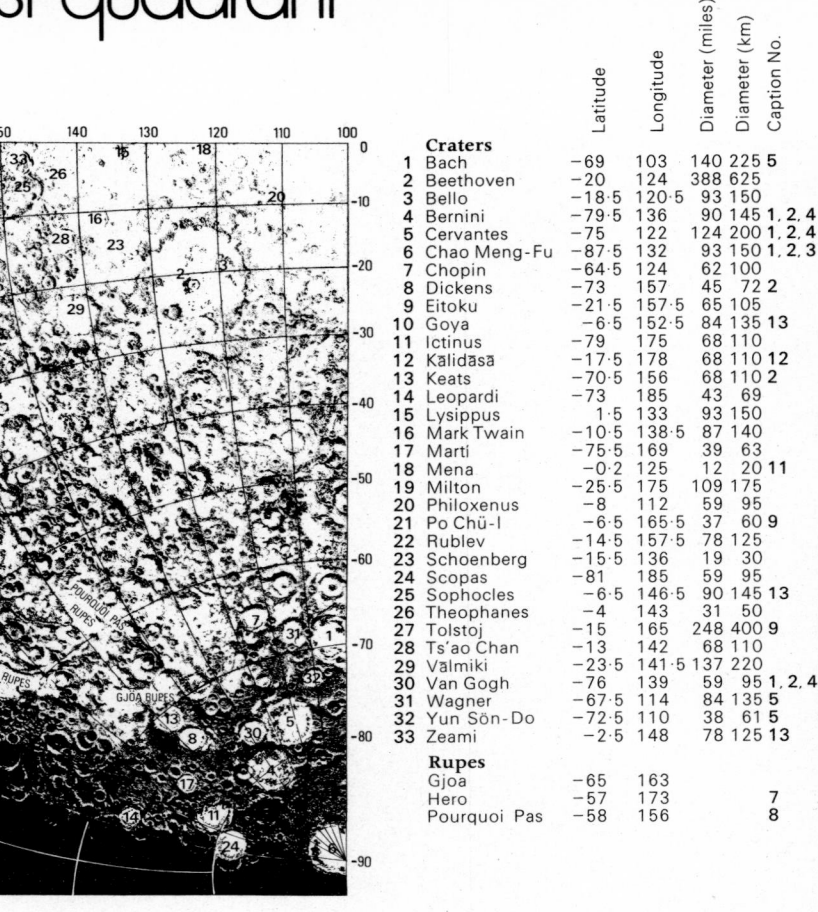

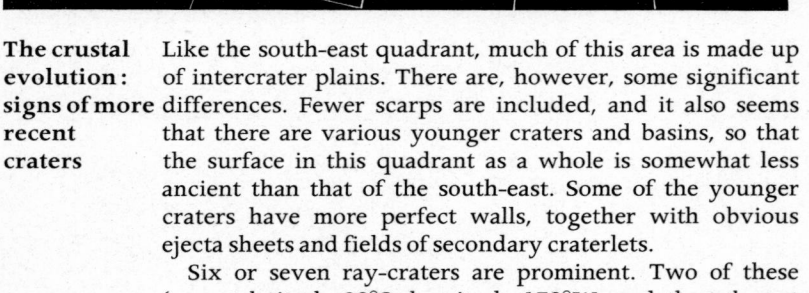

	Craters	Latitude	Longitude	Diameter (miles)	Diameter (km)	Caption No.
1	Bach	−69	103	140	225	5
2	Beethoven	−20	124	388	625	
3	Bello	−18·5	120·5	93	150	
4	Bernini	−79·5	136	90	145	1, 2, 4
5	Cervantes	−75	122	124	200	1, 2, 4
6	Chao Meng-Fu	−87·5	132	93	150	1, 2, 3
7	Chopin	−64·5	124	62	100	
8	Dickens	−73	157	45	72	2
9	Eitoku	−21·5	157·5	65	105	
10	Goya	−6·5	152·5	84	135	13
11	Ictinus	−79	175	68	110	
12	Kalidāsā	−17·5	178	68	110	12
13	Keats	−70·5	156	68	110	2
14	Leopardi	−73	185	43	69	
15	Lysippus	1·5	133	93	150	
16	Mark Twain	−10·5	138·5	87	140	
17	Marti	−75·5	169	39	63	
18	Mena	−0·2	125	12	20	11
19	Milton	−25·5	175	109	175	
20	Philoxenus	−8	112	59	95	
21	Po Chü-I	−6·5	165·5	37	60	9
22	Rublev	−14·5	157·5	78	125	
23	Schoenberg	−15·5	136	19	30	
24	Scopas	−81	185	59	95	
25	Sophocles	−6·5	146·5	90	145	13
26	Theophanes	−4	143	31	50	
27	Tolstoj	−15	165	248	400	9
28	Ts'ao Chan	−13	142	68	110	
29	Valmiki	−23·5	141·5	137	220	
30	Van Gogh	−76	139	59	95	1, 2, 4
31	Wagner	−67·5	114	84	135	5
32	Yun Sön-Do	−72·5	110	38	61	5
33	Zeami	−2·5	148	78	125	13
	Rupes					
	Gjoa	−65	163			
	Hero	−57	173			7
	Pourquoi Pas	−58	156			8

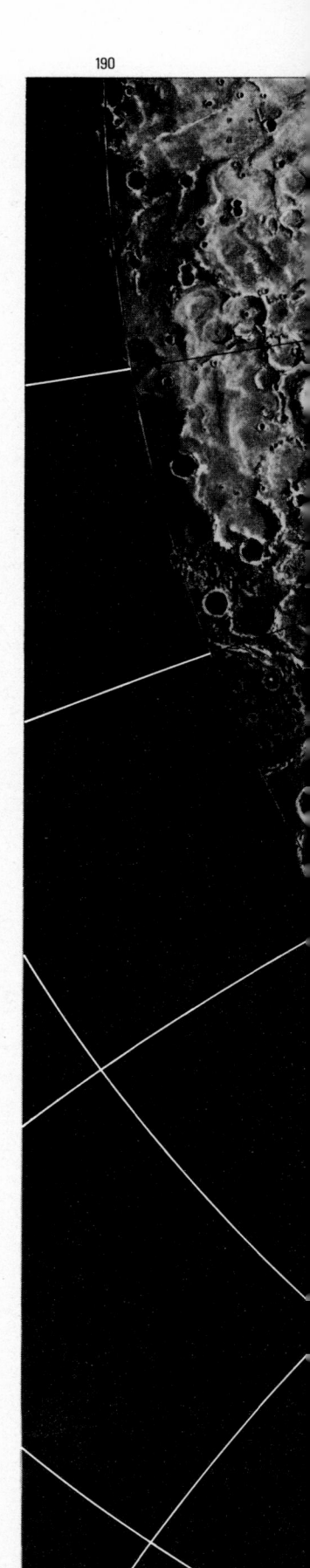

The crustal evolution: signs of more recent craters

Like the south-east quadrant, much of this area is made up of intercrater plains. There are, however, some significant differences. Fewer scarps are included, and it also seems that there are various younger craters and basins, so that the surface in this quadrant as a whole is somewhat less ancient than that of the south-east. Some of the younger craters have more perfect walls, together with obvious ejecta sheets and fields of secondary craterlets.

Six or seven ray-craters are prominent. Two of these (one at latitude 32°S, longitude 172°W, and the other at 73°S, 110°W) lie well away from the subsolar point, and are surrounded by prominent dark haloes; their light-coloured ray-systems extend for hundreds of miles.

There is one region of smooth plains material situated at the top left. This is a southern extension of Planitia Tir, most of which lies in the north-east quadrant. The other apparently smooth patch, at the top right, is the subsolar region. The area appears smoother than it really is, because the Sun is vertically above it and casts no shadows.

Beethoven— the largest crater of the south

The three largest craters have diameters greater than those of any comparable features in the south-east quadrant. The senior in size is Beethoven (2), 388 miles (625 km) wide, which has a floor flooded with smooth plains material, though it includes several features and one named crater, Bello (3). Another huge flooded crater is Tolstoj (27). The third really vast formation is as yet unnamed; it lies at latitude 45°S, longitude 177°W, and has a diameter of 250 miles (400 km). These three basins are presumably of comparable age with the large formations in the south-east quadrant. Beethoven is actually the largest enclosure on the photographed part of Mercury, apart from the Caloris Basin.

The south polar area is very rough, and contains various prominent craters, of which Cervantes (5) is typical. The polar point lies just inside the rim of Chao Meng-Fu (6), the floor of which is in permanent shadow.

The lack of named formations in wide areas of this quadrant is not significant; it merely means that the official maps of the region, being drawn up at the Jet Propulsion Laboratory in California, have not yet been finished. When they are ready, names will be allotted by the International Astronomical Union. No doubt the very large basin referred to above will be christened, as well as the ray-craters and some of the valleys—notably the long valley at 56°S, 165°W.

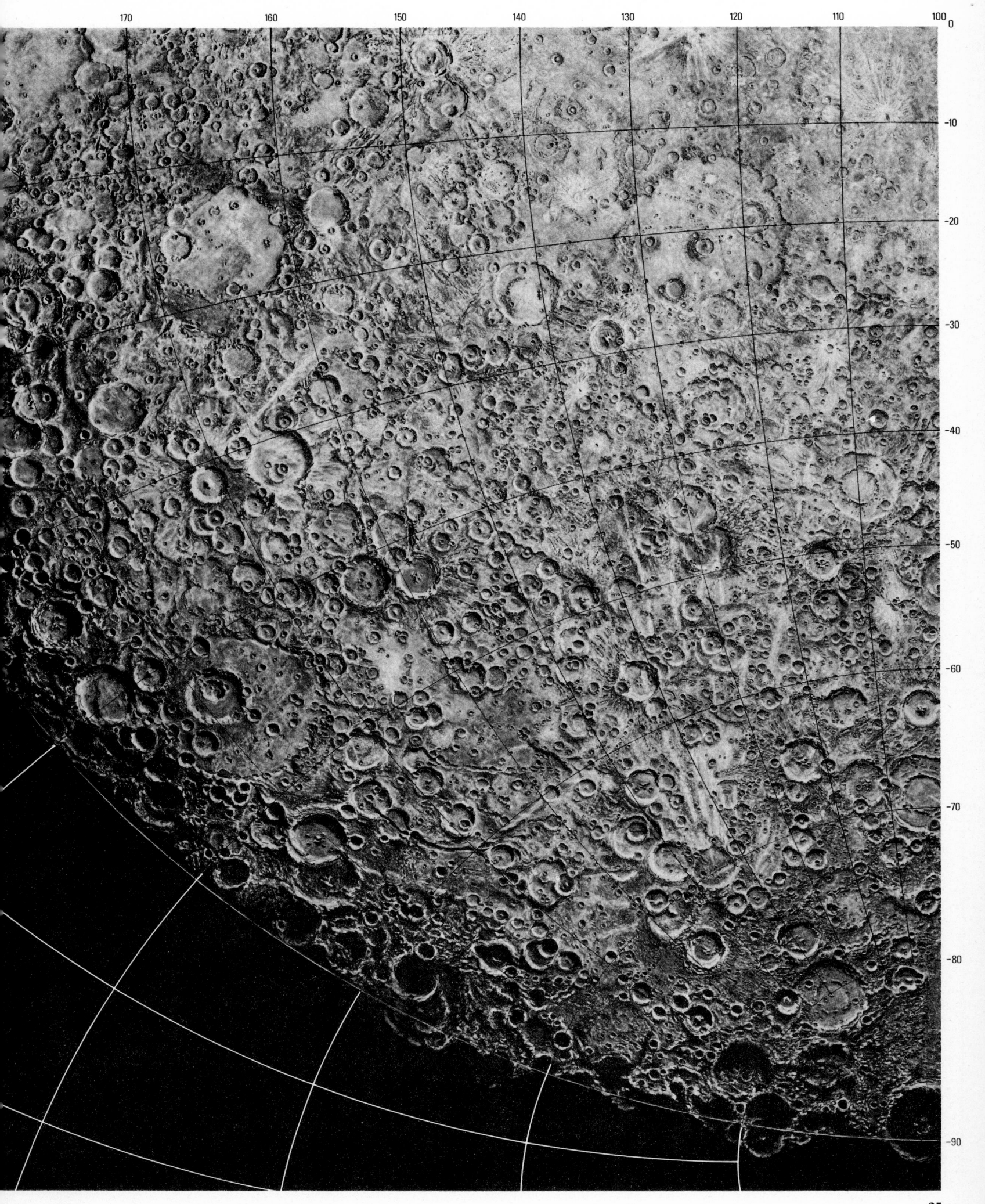

1 Bernini (A), Boccaccio (B), Cervantes (C), Chao Meng-Fu (D), Li Ching-Chao (E), Sadī (F), Van Gogh (G). Boccaccio, Li Chi'ng-Chao and Sadī are located in the south-east quadrant. The main interest centres on Chao Meng-Fu, which contains Mercury's south pole. Bernini, Cervantes and Van Gogh are shown very fore-shortened in this view.

2 Bernini (A), Cervantes (B), Chao Meng-Fu (C), Dickens (D), Keats (E), Van Gogh (F). The south polar area is again shown. Keats is interrupted by a smaller, better-formed crater. Dickens has fairly regular walls. The unnamed 30-mile (50-km) crater east of Dickens is the centre of an extensive ray-system; the rays extend over 625 miles (1,000 km) north and north-east.

3 Chao Meng-Fu (A). The south pole lies just within the crater rim; the interior is always shadowed, and must be the coldest spot on Mercury.

4 Bernini (A), Cervantes (B), Van Gogh (C). Cervantes is revealed as a very ancient double ring-wall whose west wall has been partly destroyed by the overlapping Van Gogh.

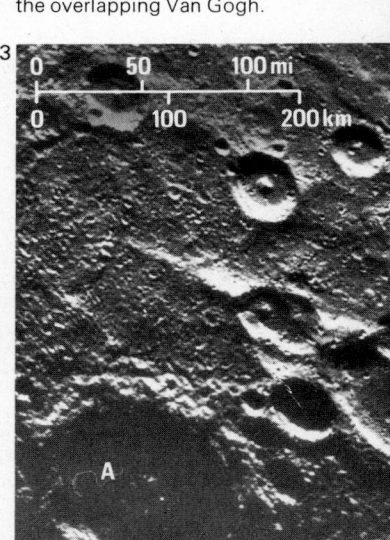

5 Bach (A), Wagner (B), Yun Sön-Do (C). The most imposing crater in this picture, Bach, has a double ring. The outer ramparts are of modest altitude, and are almost linear in places; the floor is crossed by a chain of secondary craterlets coming from Wagner. Yun Sön-Do has a floor crossed by a well-marked line of craterlets, also presumably associated with Wagner.

6 The formations in this part of Mercury have not yet been named. The centre of the picture is located at 45°S, 183°W. The most interesting feature is the large ring formation to the upper right, which has a diameter of 215 miles (350 km); it is the third largest ring so far recorded on the planet. In it is a smaller, well-marked crater, which in turn contains another crater.

7 Hero Rupes. This scarp cuts through two craters. There is a roughly parallel feature to the east, and a rather smoother narrow area to the west.

8 Pourquoi Pas Rupes. A curved scarp, the formation is less well marked than Hero Rupes, but of the same basic type.

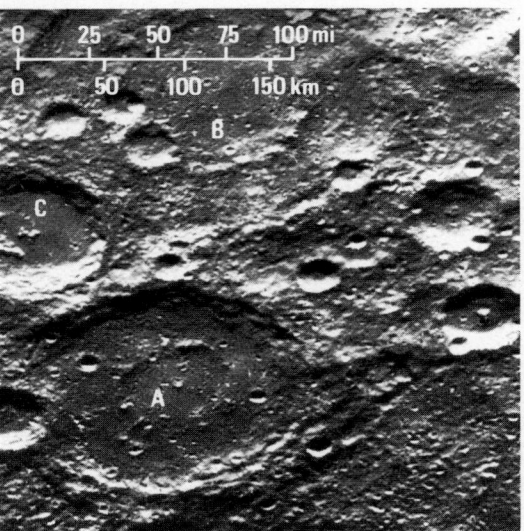

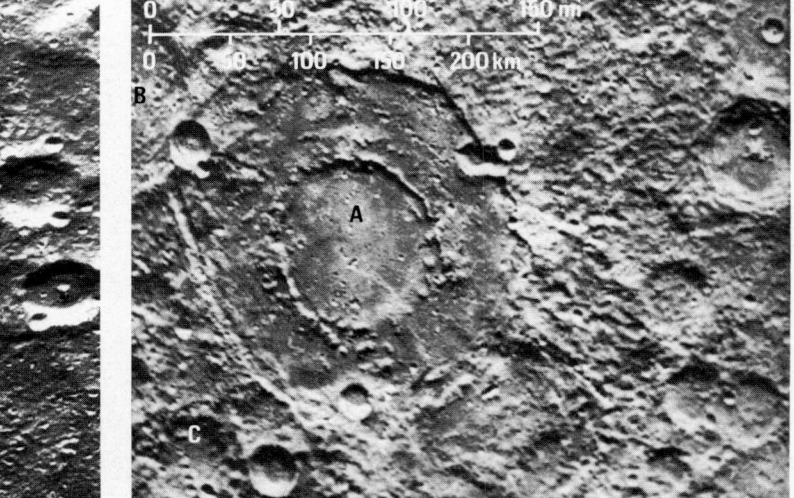

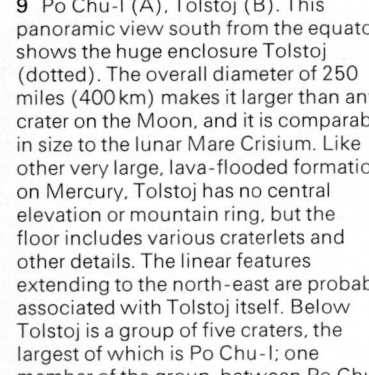

9 Po Chu-I (A), Tolstoj (B). This panoramic view south from the equator shows the huge enclosure Tolstoj (dotted). The overall diameter of 250 miles (400 km) makes it larger than any crater on the Moon, and it is comparable in size to the lunar Mare Crisium. Like other very large, lava-flooded formations on Mercury, Tolstoj has no central elevation or mountain ring, but the floor includes various craterlets and other details. The linear features extending to the north-east are probably associated with Tolstoj itself. Below Tolstoj is a group of five craters, the largest of which is Po Chu-I; one member of the group, between Po Chu-I

and the well-marked crater to its east, has been obviously deformed. To the upper right of the picture is an unnamed 55-mile (90-km) crater which is the centre of a prominent system of rays. (This picture and the detail, picture 10, are oriented with north at the bottom.)

10 The central ray-crater, as seen in 9, has regular, terraced walls and a prominent central elevation, like many ray-craters both on Mercury and on the Moon. To its north-east is an interesting group of overlapping craters; to its west, a much larger formation which contains yet another pair of overlapping craters.

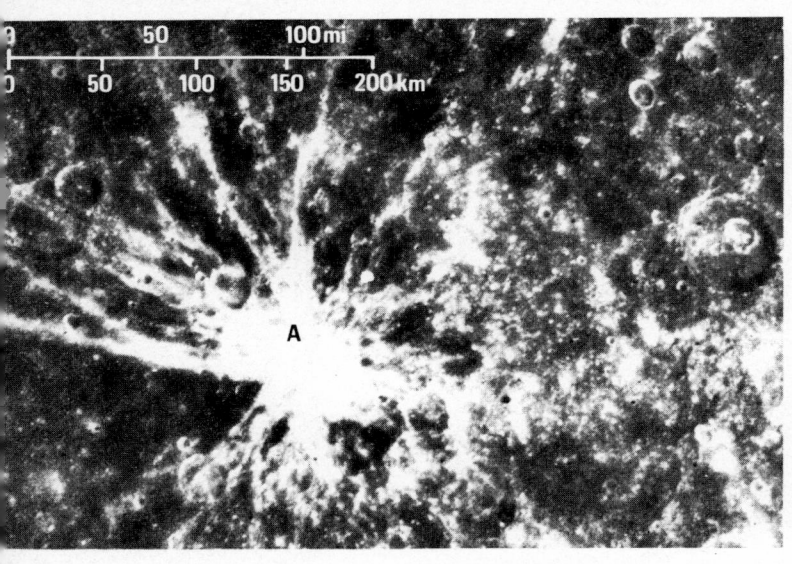

PLANITIA TIR

11 Mena (A). Though small, the crater Mena is the centre of a very prominent halo and ray-system. One interesting feature is the lack of rays in the 45-degree interval between due south and south-west, and this deficiency is certainly real, since various craterlets are shown in the ray-free area. There may be some analogy with the gap in the ray-system of the lunar crater Proclus, near the border of the Mare Crisium. Since the picture was taken under fairly high solar illumination, the Mena rays tend to mask much of the surrounding detail.

12 Kālidāsā (A), Planitia Tir. This picture, on the western terminator just south of the equator, shows the transition from the smooth, lava-flooded plains south of the Caloris Basin at the top to the densely cratered highlands at the bottom. At the top left, the lava-plains extend into Planitia Tir, located on the north-west quadrant. An interesting feature is the overlapping pair of 68-mile (110-km) craters. The western component, which is evidently the younger, has been named Kālidāsā. Nearly the western wall of its older neighbour has survived the formation of Kālidāsā, remaining as a modified central peak.

13 Goya (A), Sophocles (B), Zeami (C). The most recent crater in this region is Zeami, with a multiple central peak and strongly terraced walls. Beyond Zeami is an extensive field of secondary craterlets, overlying the older crater Sophocles.

29

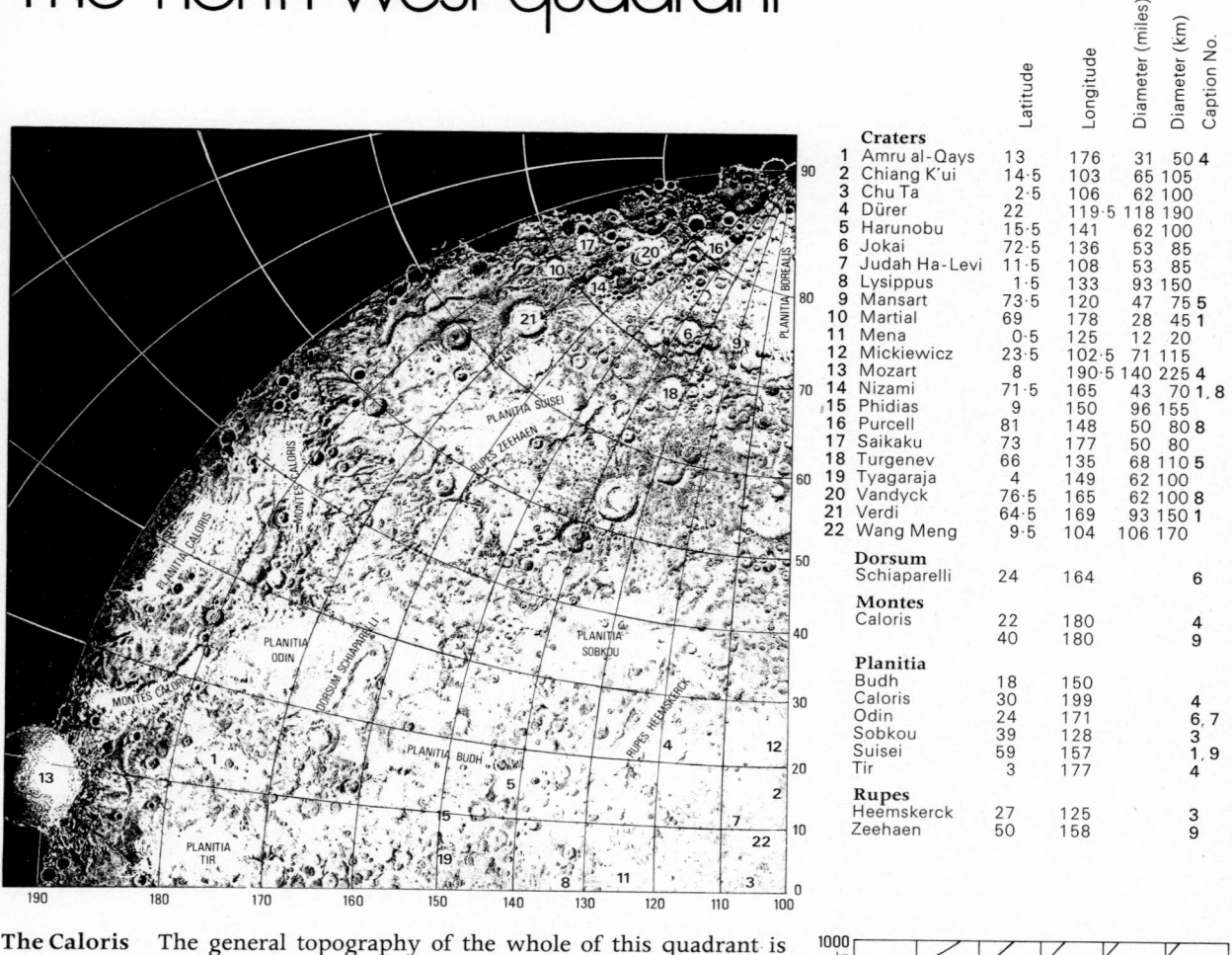

	Craters	Latitude	Longitude	Diameter (miles)	Diameter (km)	Caption No.
1	Amru al-Qays	13	176	31	50	4
2	Chiang K'ui	14·5	103	65	105	
3	Chu Ta	2·5	106	62	100	
4	Dürer	22	119·5	118	190	
5	Harunobu	15·5	141	62	100	
6	Jokai	72·5	136	53	85	
7	Judah Ha-Levi	11·5	108	53	85	
8	Lysippus	1·5	133	93	150	
9	Mansart	73·5	120	47	75	5
10	Martial	69	178	28	45	1
11	Mena	0·5	125	12	20	
12	Mickiewicz	23·5	102·5	71	115	
13	Mozart	8	190·5	140	225	4
14	Nizami	71·5	165	43	70	1, 8
15	Phidias	9	150	96	155	
16	Purcell	81	148	50	80	8
17	Saikaku	73	177	50	80	
18	Turgenev	66	135	68	110	5
19	Tyagaraja	4	149	62	100	
20	Vandyck	76·5	165	62	100	8
21	Verdi	64·5	169	93	150	1
22	Wang Meng	9·5	104	106	170	

Dorsum					
Schiaparelli	24	164			6

Montes					
Caloris	22	180			4
	40	180			9

Planitia					
Budh	18	150			4
Caloris	30	199			4
Odin	24	171			6, 7
Sobkou	39	128			3
Suisei	59	157			1, 9
Tir	3	177			4

Rupes					
Heemskerck	27	125			3
Zeehaen	50	158			9

The Caloris Basin—beyond into the uncharted

The general topography of the whole of this quadrant is dominated by the Caloris Basin, part of which is shown. Unfortunately the centre lies beyond the terminator, at latitude 30°N, longitude 190°W, and was not photographed.

The Caloris Basin is estimated at 840 miles (1,350 km) in diameter, and is therefore comparable in size with the lunar Mare Imbrium. The resemblance may be more than superficial. There is no doubt that the formation of the Mare Imbrium had profound effects over much of the lunar surface, and the same may well be true of the Caloris Basin.

Adjacent areas affected by the Caloris event

Outside the mountain ring lies a complex terrain of ancient craters and basins which have been strongly sculptured by shallow valleys radial to the Caloris Basin. Many areas have also been flooded by the smooth plains material, but others are covered with rougher material—making up a terrain of hummocky plains. This area extends over 600 miles (1,000 km) around Caloris, and it contains five of the six other plains so far named—Tir, Budh, Odin, Sobkou and Suisei. The effects of the Caloris event die away with increasing distance, and there is a steady gradation back into the intercrater plain kind of terrain which is so characteristic of the other quadrants.

In some regions the older structures have been completely obliterated (as also happens round the lunar Mare Imbrium), but here and there, especially round the margins of the smooth plains material, ghost craters may be seen. Investigators have a good opportunity to study the younger, smaller formations in this region, since the confusing background of older structures has been largely destroyed.

Ray-craters and the post-Caloris Mozart crater

There are four prominent ray-craters in this quadrant, counting the close pair in Planitia Sobkou as a single centre. The map shows that the peculiar curved ray seen to the south and east of these in the more distant views breaks down into straight segments in the map projection. The very prominent component running north-north-east is clearly a northward extension of Heemskerck Rupes.

The large crater Mozart (13) is of special interest. It was obviously formed after the Caloris Basin, and must be one of the youngest of the main craters. Its west wall is the illuminated area which reached farthest into the night hemisphere at the times when Mariner 10 flew past Mercury.

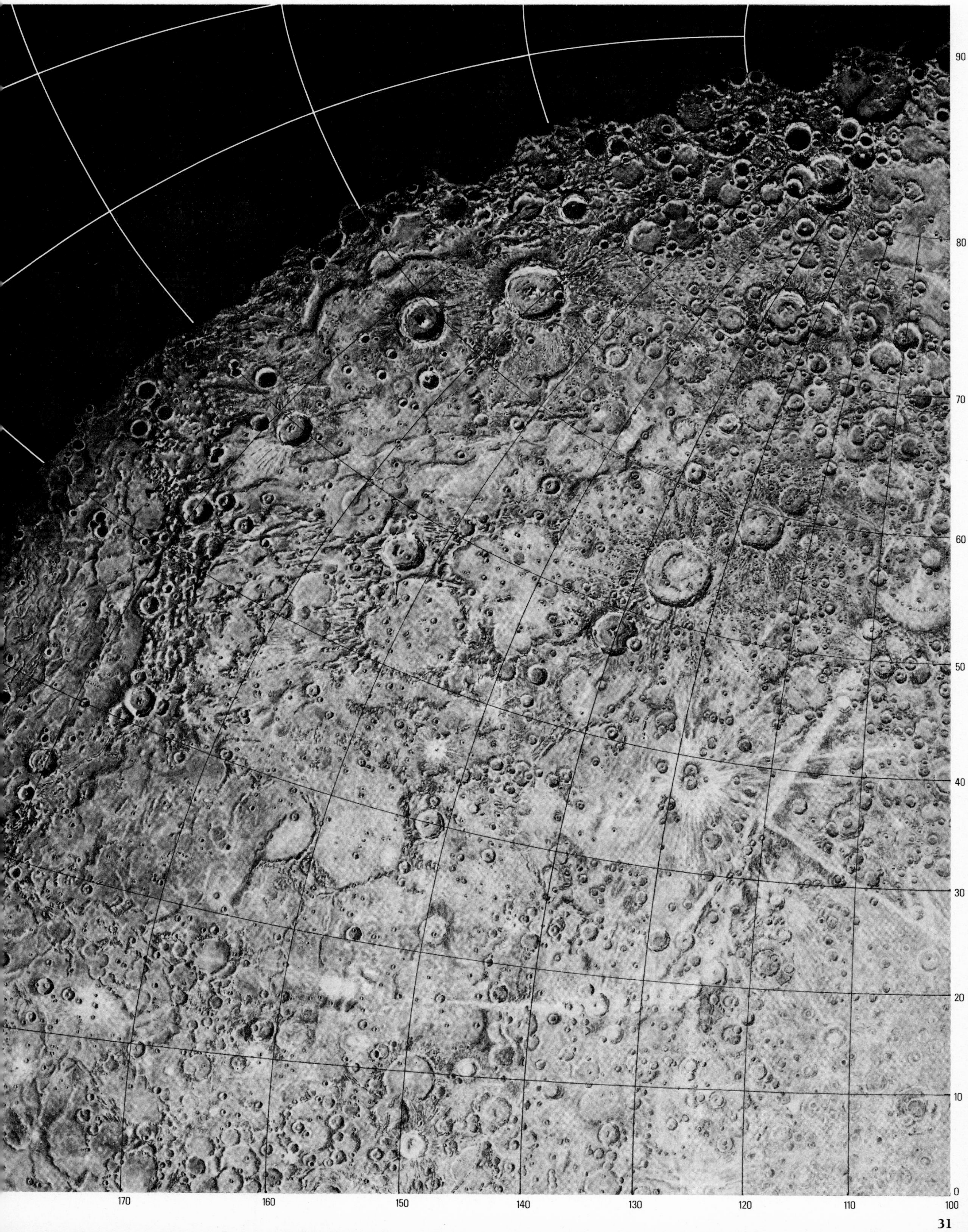
90
80
70
60
50
40
30
20
10
0
170
160
150
140
130
120
110
100

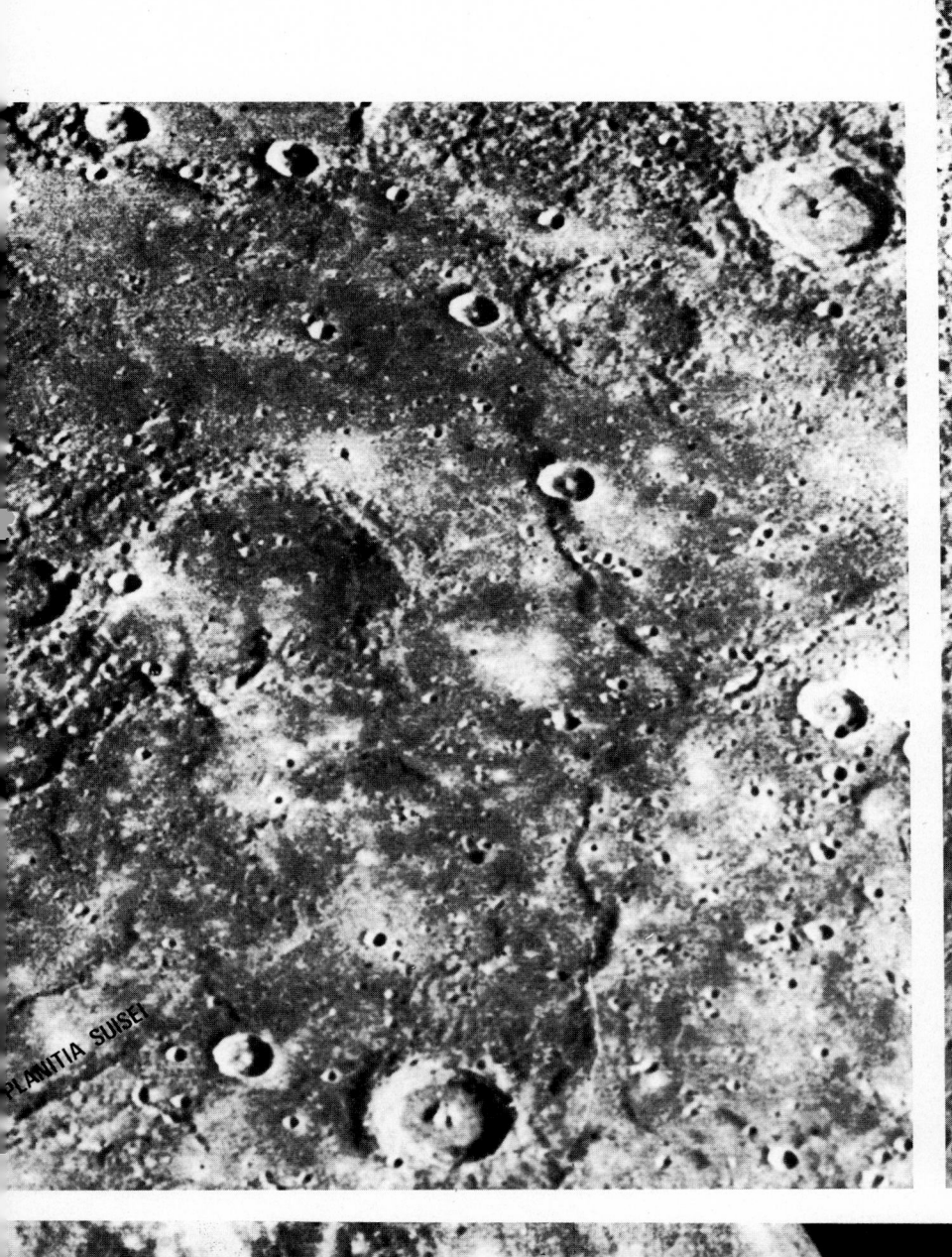

| 0 | 25 | 50 | 75 | 100 | 125 mi |
| 0 | 50 | 100 | 150 | 200 km |

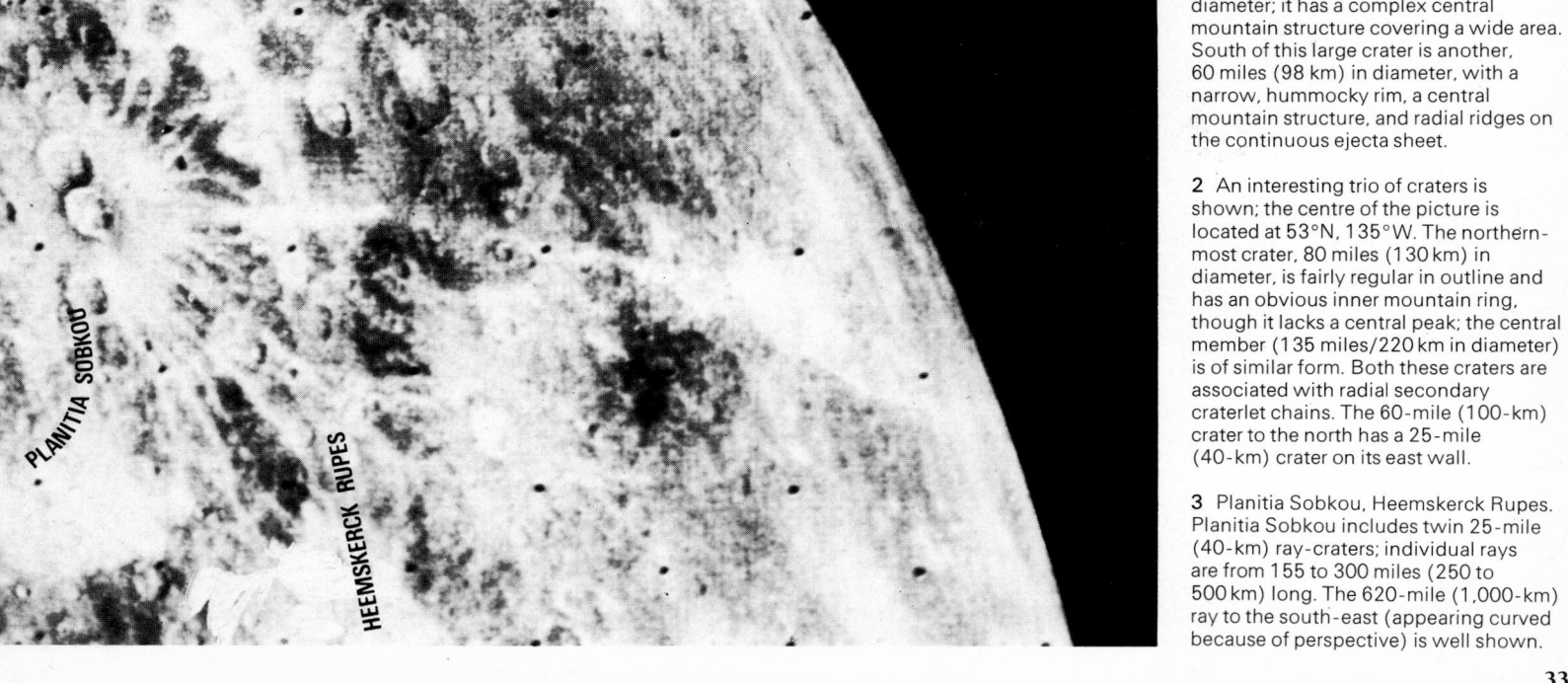

PLANITIA SUISEI

PLANITIA SOBKOU

HEEMSKERCK RUPES

1 Martial (A), Nizami (B), Verdi (C), Planitia Suisei. The prominent crater to the upper left is 85 miles (137 km) in diameter; it has a complex central mountain structure covering a wide area. South of this large crater is another, 60 miles (98 km) in diameter, with a narrow, hummocky rim, a central mountain structure, and radial ridges on the continuous ejecta sheet.

2 An interesting trio of craters is shown; the centre of the picture is located at 53°N, 135°W. The northernmost crater, 80 miles (130 km) in diameter, is fairly regular in outline and has an obvious inner mountain ring, though it lacks a central peak; the central member (135 miles/220 km in diameter) is of similar form. Both these craters are associated with radial secondary craterlet chains. The 60-mile (100-km) crater to the north has a 25-mile (40-km) crater on its east wall.

3 Planitia Sobkou, Heemskerck Rupes. Planitia Sobkou includes twin 25-mile (40-km) ray-craters; individual rays are from 155 to 300 miles (250 to 500 km) long. The 620-mile (1,000-km) ray to the south-east (appearing curved because of perspective) is well shown.

5 Mansart (A), Turgenev (B), Planitia Borealis. The main part of this region is very rough, but it adjoins the much smoother Planitia Borealis located on the north-east quadrant. The area is clearly representative of a more heavily cratered surface. Mansart, with a well-defined central peak, encroaches on a formation of similar size. However the overlap is slight when compared to the more excessive examples such as that of Kalidasa and its counterpart on the south-west quadrant (picture 12). Such overlapping is hard to explain by any conventional theory of crater formation, particularly based on studies of the surface features of the Moon.

4 Amru al-Qays (A), Mozart (B), Montes Caloris, Planitia Caloris, Planitia Tir. The Montes Caloris form an irregular blocky scarp standing some 1¼ miles (2 km) above the edge of the plain. The whole area between here and the small crater Amru al-Qays has been obviously affected by the event which formed the Caloris Basin (Planitia Caloris). Another smooth and sparsely cratered lava plain is Planitia Tir, which is crossed by the equator. On some pictures the west wall of the crater Mozart is faintly illuminated; this illuminated rim projects farther into the dark hemisphere than any other formations around the terminator.

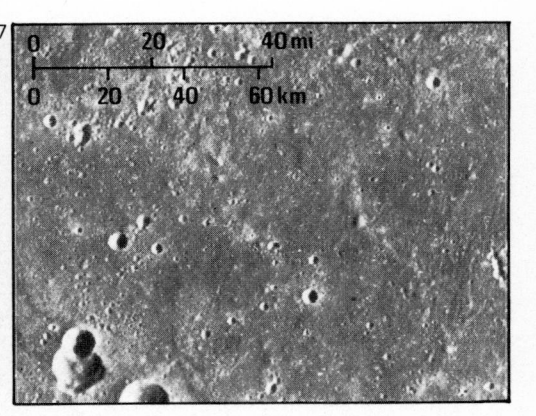

PLANITIA
ODIN

DORSUM SCHIAPARELLI

6 Planitia Odin, Dorsum Schiaparelli. The Planitia
Odin is a relatively smooth plain; on the west it
slopes down into Planitia Caloris, where it interrupts
the Montes Caloris chain. It is bounded to the east
by Dorsum Schiaparelli, whose north end bisects a
150-mile (240-km) lava-flooded crater.

7 In this close-up the whole area can be seen
to contain many craterlets and ridges probably
associated with the Planitia Caloris development.

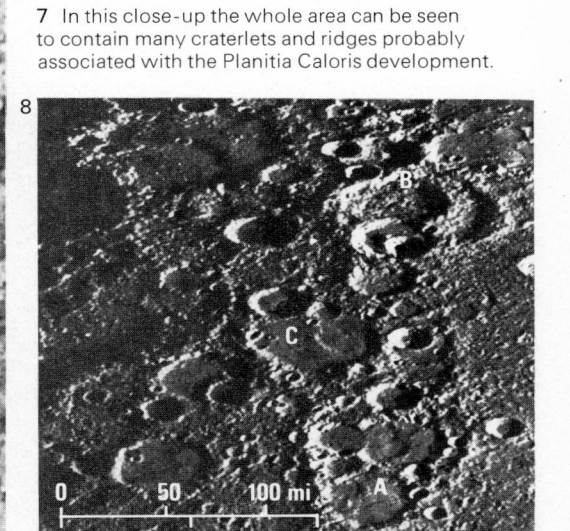

8 Nizami (A), Purcell (B), Vandyck (C). This is a closely
cratered group in the north polar region.

9 Planitia Suisei, Rupes Zeehaen. Planitia Suisei is another
of the relatively smooth lava-plains. Rupes Zeehaen appears to
be formed from the remnants of walls of craters destroyed in the
lava-floods following the formation of Planitia Caloris.

10 Lineated terrain, centred on 43°N, 164°W, is evident outside
the north-east edge of the Caloris mountain ring.

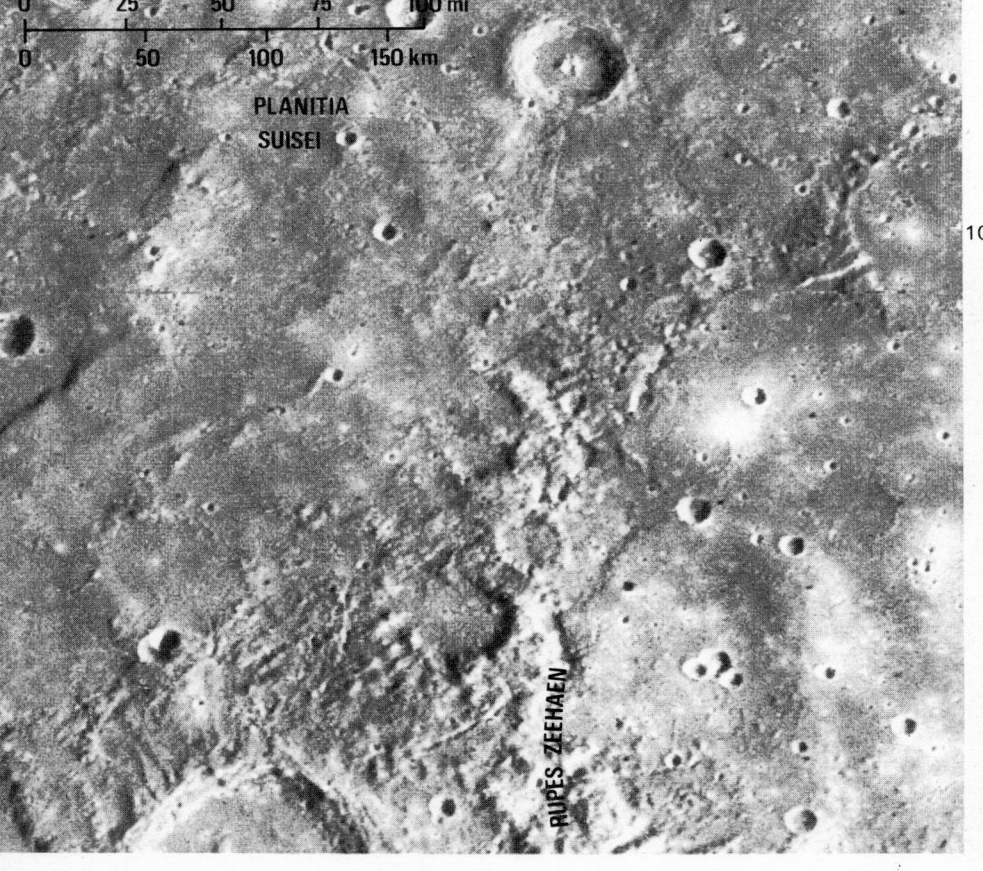

PLANITIA
SUISEI

RUPES ZEEHAEN

The north-east quadrant

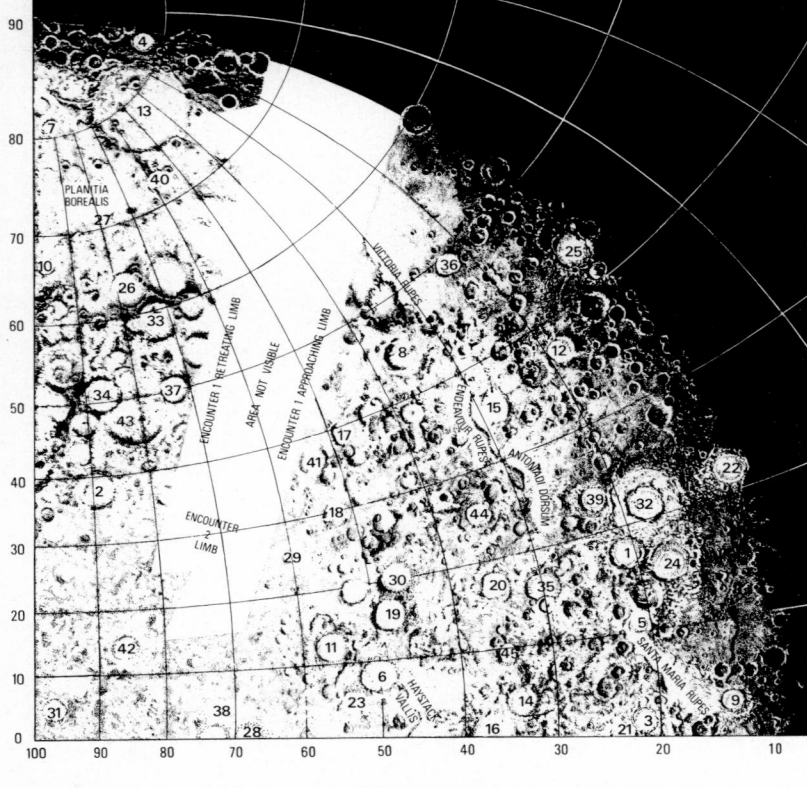

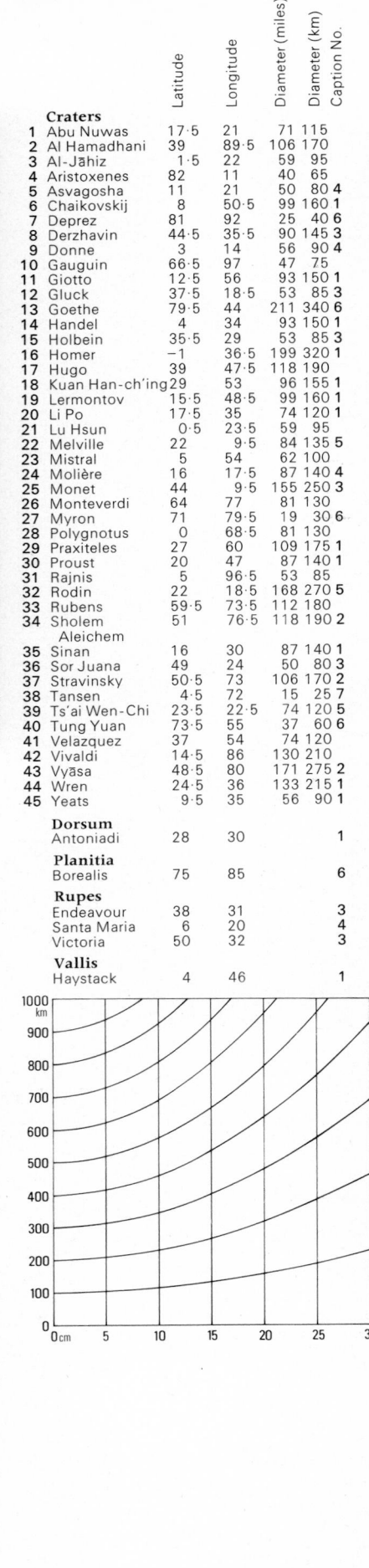

		Latitude	Longitude	Diameter (miles)	Diameter (km)	Caption No.
Craters						
1	Abu Nuwas	17·5	21	71	115	
2	Al Hamadhani	39	89·5	106	170	
3	Al-Jāhiz	1·5	22	59	95	
4	Aristoxenes	82	11	40	65	
5	Asvagosha	11	21	50	80	4
6	Chaikovskij	8	50·5	99	160	1
7	Deprez	81	92	25	40	6
8	Derzhavin	44·5	35·5	90	145	3
9	Donne	3	14	56	90	4
10	Gauguin	66·5	97	47	75	
11	Giotto	12·5	56	93	150	1
12	Gluck	37·5	18·5	53	85	3
13	Goethe	79·5	44	211	340	6
14	Handel	4	34	93	150	1
15	Holbein	35·5	29	53	85	3
16	Homer	−1	36·5	199	320	1
17	Hugo	39	47·5	118	190	
18	Kuan Han-ch'ing	29	53	96	155	1
19	Lermontov	15·5	48·5	99	160	1
20	Li Po	17·5	35	74	120	1
21	Lu Hsun	0·5	23·5	59	95	
22	Melville	22	9·5	84	135	5
23	Mistral	5	54	62	100	
24	Molière	16	17·5	87	140	4
25	Monet	44	9·5	155	250	3
26	Monteverdi	64	77	81	130	
27	Myron	71	79·5	19	30	6
28	Polygnotus	0	68·5	81	130	
29	Praxiteles	27	60	109	175	1
30	Proust	20	47	87	140	1
31	Rajnis	5	96·5	53	85	
32	Rodin	22	18·5	168	270	5
33	Rubens	59·5	73·5	112	180	
34	Sholem Aleichem	51	76·5	118	190	2
35	Sinan	16	30	87	140	1
36	Sor Juana	49	24	50	80	3
37	Stravinsky	50·5	73	106	170	2
38	Tansen	4·5	72	15	25	7
39	Ts'ai Wen-Chi	23·5	22·5	74	120	5
40	Tung Yuan	73·5	55	37	60	6
41	Velazquez	37	54	74	120	
42	Vivaldi	14·5	86	130	210	
43	Vyāsa	48·5	80	171	275	2
44	Wren	24·5	36	133	215	1
45	Yeats	9·5	35	56	90	1
Dorsum						
	Antoniadi	28	30			1
Planitia						
	Borealis	75	85			6
Rupes						
	Endeavour	38	31			3
	Santa Maria	6	20			4
	Victoria	50	32			3
Vallis						
	Haystack	4	46			1

The areas that Mariner could not photograph

Because of the flight path of Mariner 10, both the strip of surface on this quadrant and the whole of the reverse hemisphere were inaccessible—and this applied during each of the three passes during which information could be sent back. Mariner 10 is now dead, and to complete the coverage a new vehicle would either have to pass Mercury at a different time in the Mercurian year or else enter a closed path around the planet—as Mariner 9 and the two Viking orbiters have already done with Mars. However, there is no reason to believe that the unexamined area is basically different from that which has been covered.

Unfortunately, the regions to either side of the missing strip appeared highly foreshortened in the only views that Mariner 10 was able to obtain, so that analysis of them is far from easy. There is some smooth plains material (Planitia Borealis) and a large lava-flooded basin, 190 miles (310 km) in diameter, lying between the Planitia and the pole. This lunar-type region is completely separated from the much more extensive plains near the Caloris Basin by a belt of heavily cratered terrain.

Ray-craters and the effect of the Sun

There are seven prominent ray-craters in this quadrant. These appear to be concentrated towards the subsolar region, but this may be partly misleading, because rays appear only under high illumination—and ray-craters on Mercury are thus easier to detect when the Sun is almost above them.

Of the more interesting features, one is Rodin (32), a particularly large and well-developed double ring basin. Its associated field of ejecta and secondary cratering is very well displayed, because it was close to the terminator when photographed, and it lies in a relatively smooth area; around it are other major craters, such as Molière (24), Abu Nuwas (1) and the smaller but well-formed Ts'ai Wen-Chi (39). Right on the terminator, with a shadowed floor, is yet another well-formed crater, Melville (22). Of the four named valleys, one (Haystack) is a crater-chain. Also in the quadrant is one of the two named ridges, Dorsum Antoniadi, which seems to be a southern extension of Endeavour Rupes. It would appear that the scarp has turned into a ridge simply because the level on the upthrust side of the fault has declined more abruptly. Equally it has been suggested that Goethe (13) is geologically related to the Planitia Borealis.

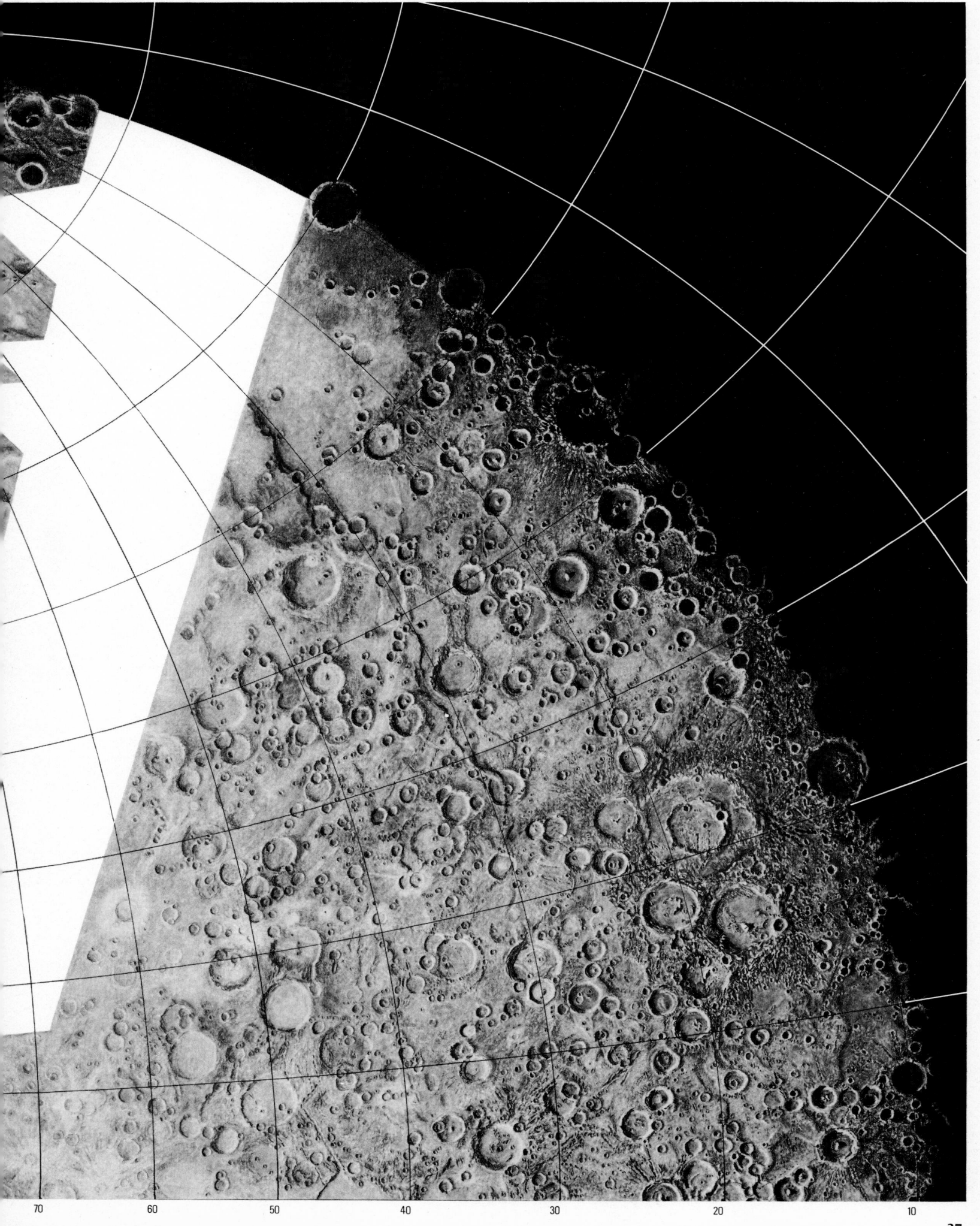

70   60   50   40   30   20   10

1 Chaikovskij (A), Giotto (B), Handel (C), Homer (D), Kuan Han-ch'ing (E), Lermontov (F), Li Po (G), Praxiteles (H), Proust (I), Sinan (J), Wren (K), Yeats (L), Dorsum Antoniadi, Haystack Vallis. These views were obtained during the first encounter approach; to the upper left, much of the topographical detail is obscured because of the high Sun angle. To the north is the Dorsum Antoniadi, one of the two major named ridges on Mercury, which cuts right through a well-formed crater; it extends farther north, where it assumes the characteristics of a scarp. Sinan is somewhat deformed, and has its south wall broken by younger craters; Li Po is notable because of the peculiar spiral chain of craterlets winding outwards to its east and south. Lermontov has a white floor, and any detail is obscured; it is clearly younger than its neighbour Proust. Giotto is reasonably well formed, but again no detail can be seen inside it under these conditions of illumination. Homer has a relatively smooth floor apart from one considerable crater and some smaller pits. The Haystack Vallis is a chain of small craterlets rather than a true valley.

2 Sholem Aleichem (A), Stravinsky (B) and Vyāsa (C) (dotted). The region immediately beyond the limb to the upper right has not been photographed at all, so that coverage of this quadrant remains incomplete. To the top centre of the picture, not far from the limit of the photographed zone, is a prominent crater 75 miles (120 km) in diameter; below this is a larger formation with a crater-pitted floor but no central peak; and to the left is an even larger enclosure, more than 185 miles (300 km) across, which has a chain of craterlets and peaks across its floor. Only a small part of the eastern wall of this formation appears on the north-east quadrant map.

3 Derzhavin (A), Gluck (B), Holbein (C), Monet (D), Sor Juana (E), Endeavour Rupes, Victoria Rupes. This one mosaic provides all the information available for mapping the north-east quadrant north of latitude 25° and east of longitude 40°. It shows two of the major scarps on Mercury: Victoria Rupes and the northern half of the Endeavour Rupes. The height of the Victoria Rupes is approximately 1¼ miles (2 km). In the centre of the picture is a well-formed crater 60 miles (100 km) in diameter with a prominent single central peak; almost all central elevations in craters larger than this are complex.

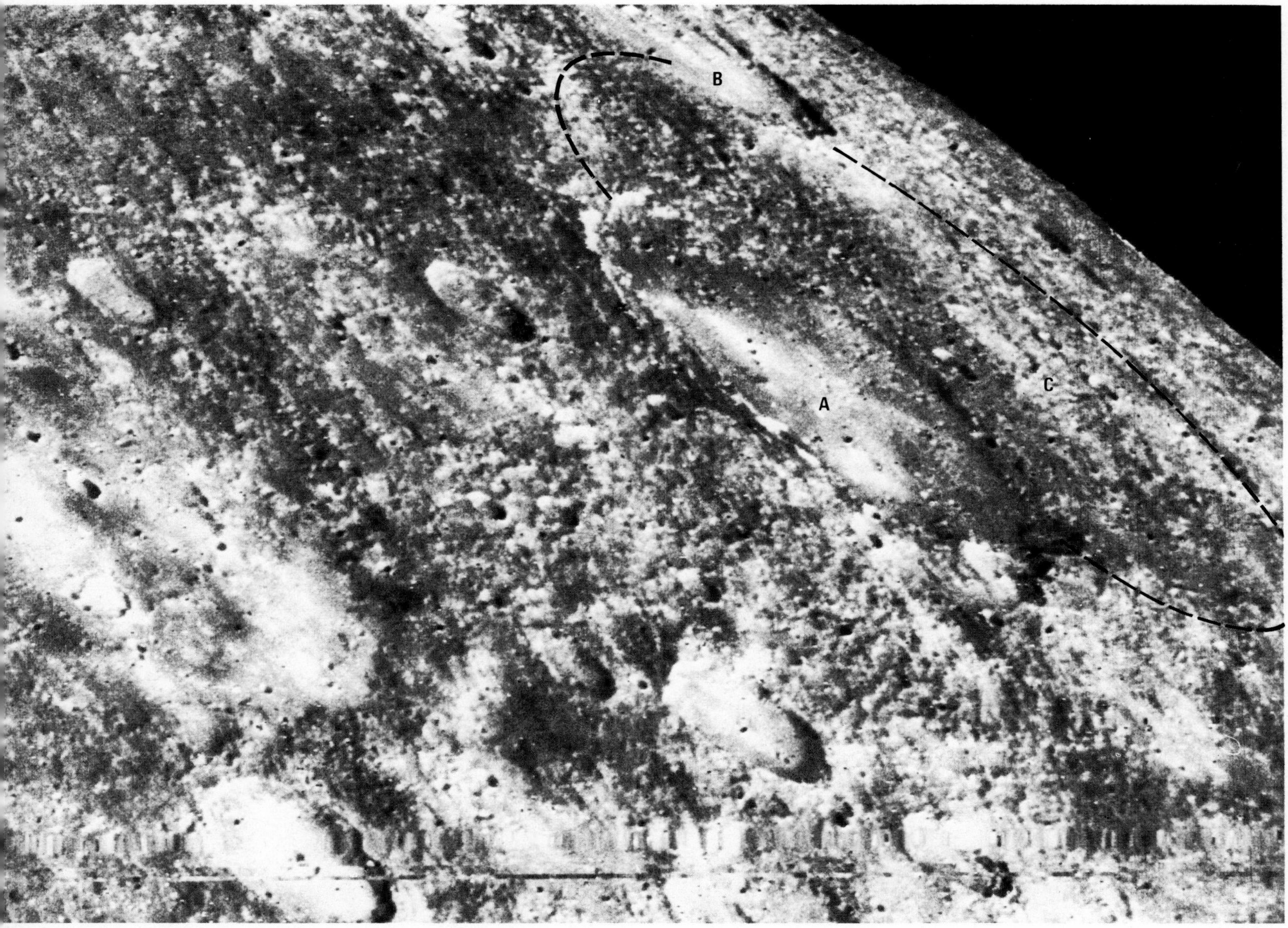

4 Asvagosha (A), Donne (B), Molière (C), Santa Maria Rupes. A prominent feature is the chain of five well-formed craters. Molière, the largest of the chain, has an elevation slightly away from the centre, lying on the edge of what appears to be a small ring. The crater between Molière and Asvagosha seems to be older, with a less regular rampart and a rougher floor. The Santa Maria Rupes leads southwards from Asvagosha to the peculiar multiple crater almost due west of Donne, which is very regular, with high terraced walls.

5 Melville (A), Rodin (B), Ts'ai Wen-Chi (C). Rodin is a double ring-wall 165 miles (270 km) in diameter; the main crater is surrounded by a concentric outer ring, which extends as far as Ts'ai Wen-Chi. The walls of Rodin itself are comparatively low, and are in places broken by craterlets; the floor contains many small pits. Ts'ai Wen-Chi is better defined than Rodin; the walls are higher and show some terracing, and there is a central elevation. The area above Rodin and Ts'ai Wen-Chi is comparatively smooth.

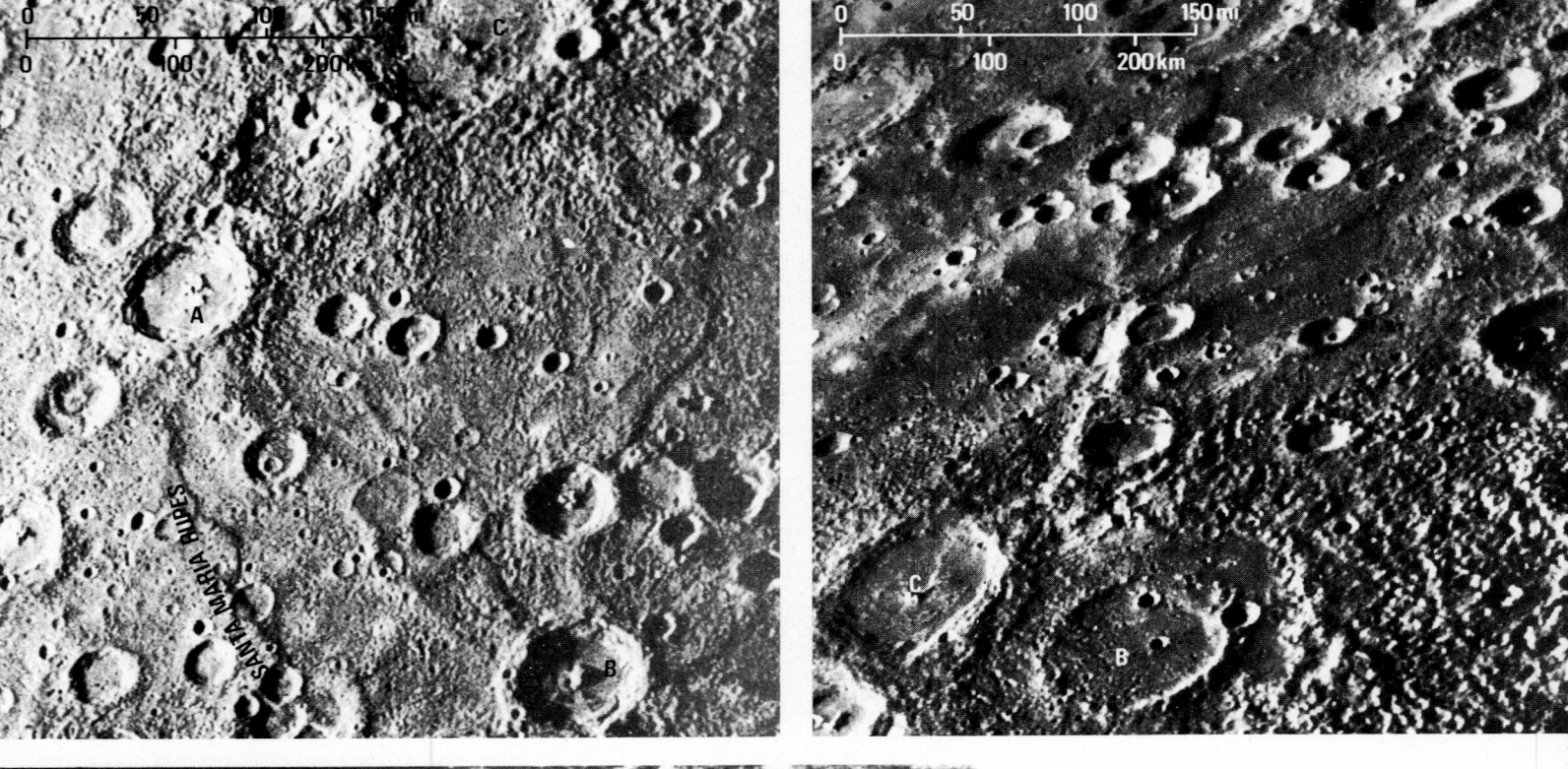

6 Deprez (A), Goethe (B), Jokai (C), Mansart (D), Myron (E), Purcell (F), Tung Yuan (G), Planitia Borealis. Jokai, Mansart and Purcell are located on the north-west quadrant. There is an obvious gradation from the rough polar area, in the lower left-hand corner, to the relative smoothness of Planitia Borealis.

Handel (A). This close-up, taken in the final stages of the first encounter approach sequence illustrates the closely pitted surface associated with the only named crater, Handel, and a similar formation about one-third its diameter. This rugged terrain is the result of a high concentration of secondary craterlets. Similar rugged terrain is characteristic of much of the intercrater plains in the rest of the north-east quadrant. The western two-thirds of Handel is visible in picture 1 which just shows the edge of the crater pair on its north-east wall. Most of the small craters around Handel have clearly arisen as secondaries from Handel itself. Many of them are oriented in chains pointing away from the main crater and their association is clearly evident.

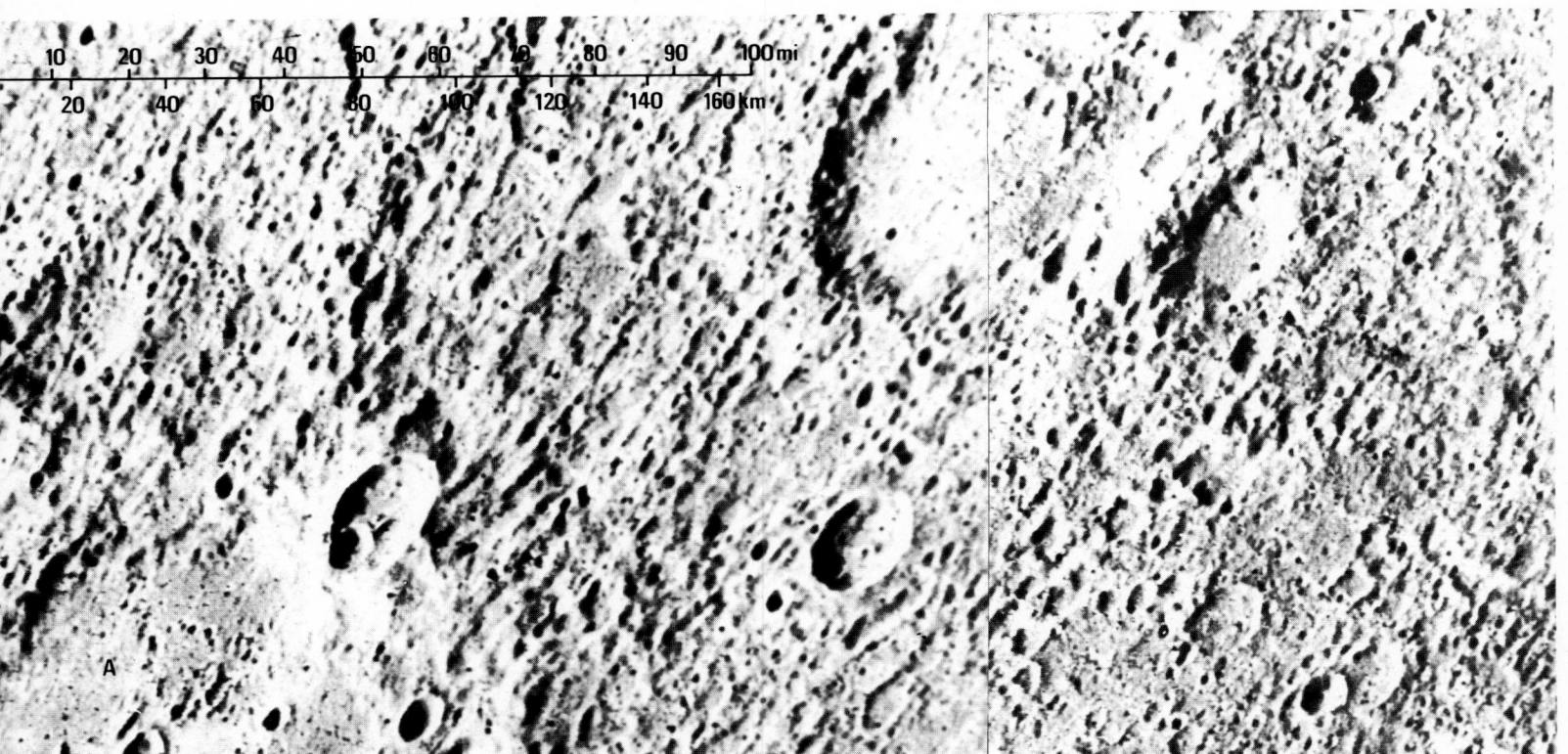

PLANITIA BOREALIS

8 Boethius (A), Tansen (B). This view was obtained under high solar illumination. There is a prominent ray-crater, Tansen, and a much larger double ring-wall, Beothius (dotted; located on the south-east quadrant), which is practically invisible under these conditions.

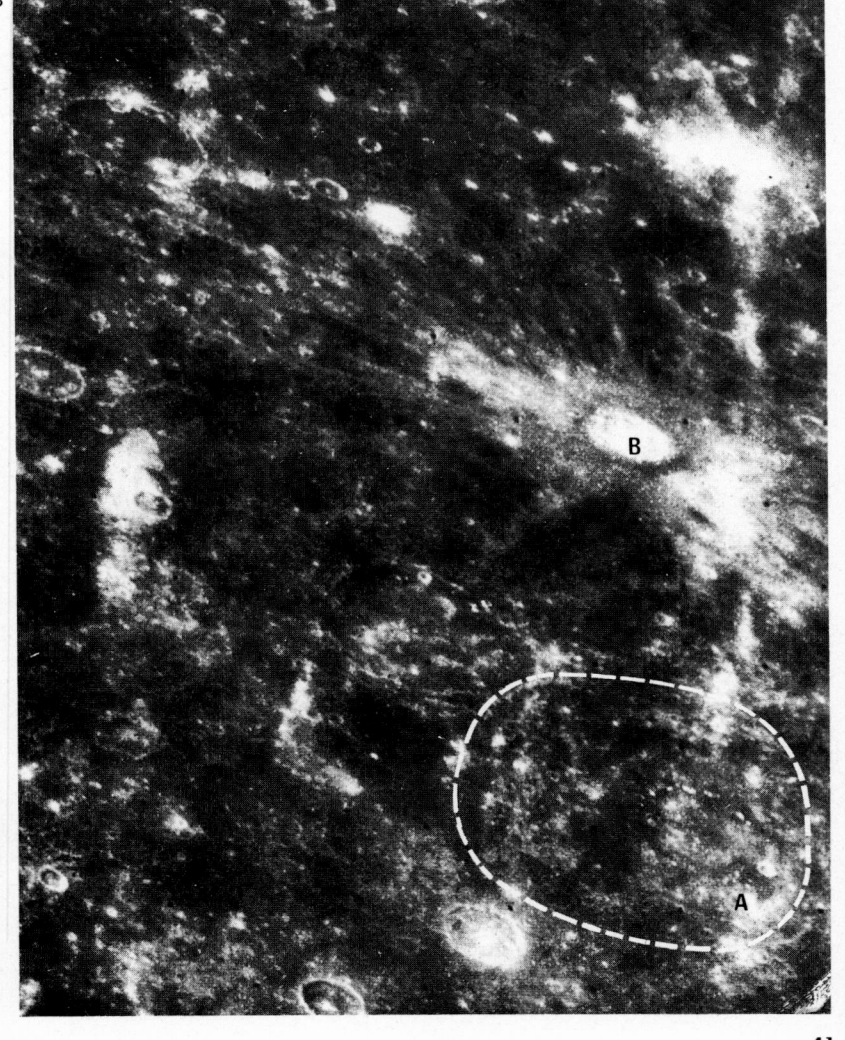

Conditions on the surface

Similarities with the Moon's topography

Superficially, the surface of Mercury looks very like that of the Moon. There are some differences in detail, but it seems fairly certain that the same forces moulded the surface features of the two worlds. It has been established that the Earth is between 4,500 and 5,000 million years old—or four and a half and five aeons (one aeon being equal to 1,000 million years). Samples brought back from the Moon by the Apollo astronauts and the Russian automatic probes confirm that the lunar age is about the same, and we may assume this also for Mercury.

If the origin of the craters and other features of the Moon could be established, it would be possible to decide about those of Mercury, but the situation is by no means clear-cut. The main controversy centres on whether the Moon's craters were produced by volcanic activity, meteoritic bombardment producing impact structures, or a combination of both. In all probability the latter is true—both volcanic and impact craters exist on the Moon, and presumably do on Mercury as well.

The impact and volcanic theories

According to the impact theory, Mercury initially developed a comparatively firm crust covering the inner magma. Using the parallel of the surface history of the Moon, it can be assumed that intense meteoritic bombardment produced huge basins on Mercury more than four aeons ago. About 3·9 aeons ago came the greatest impact of all, which produced the vast Caloris Basin—probably the counterpart of the impact that created the Mare Imbrium on the Moon. Tremendous lava-flows had already started to pour forth from inside the planet, filling the basins, and this activity continued until 3·1 aeons ago, when it stopped rather abruptly. Meanwhile, smaller meteoritic falls had produced craters and walled plains, some of which were filled with lunar-type *mare* material. As the lava outpouring slowed down, the more recent craters were left relatively undamaged and the youngest of them unfilled, with their central mountain structures clear-cut. Ray-craters would be the youngest of all the craters, since the bright streaks emanating from them cross all other formations.

On the volcanic theory, the main craters, as well as the regular plains, may be in the nature of calderas. (A caldera is a volcanic crater formed by the collapse of the surface into an underground cavity.) The time-scale is of the same order as for the impact theory, although many more recent meteoritic craters must exist.

Although Mariner 10 has shown that the laws of crater distribution and arrangement on Mercury are essentially lunar, there are structural differences between Mercury and the Moon that may well be associated with the greater surface gravity of the former. One distinction has been suggested: the Mercurian intercrater plains may represent a very ancient surface, whose lunar analogue has been largely obliterated by the subsequent formation of more numerous large basins and craters that make up the regular *maria*.

Temperature variations of Moon and Mercury

Although Mercury has the harshest surface environment of any planet in the Solar System, the thermal properties of the surface material appear to be identical to those of the Moon. At its closest approach to the Sun, Mercury receives ten times as much solar energy as the Moon. At the equator, temperatures can soar to 800°F (425°C) and in the dark hemisphere the surface rapidly cools to less than −275°F (−170°C). However, Mercury's thermal inertia—that is, the time it takes for the surface to absorb or conduct heat—is very similar to the Moon's. Thus both planets cool and heat up very rapidly. The only distinction between the thermal properties of the two planets is defined by Mercury's close proximity to the Sun; thus, unlike the Moon, where the subsurface temperatures are always below freezing over the whole globe, on Mercury the subsurface temperatures are always above freezing at the equator, although well below freezing in the polar regions.

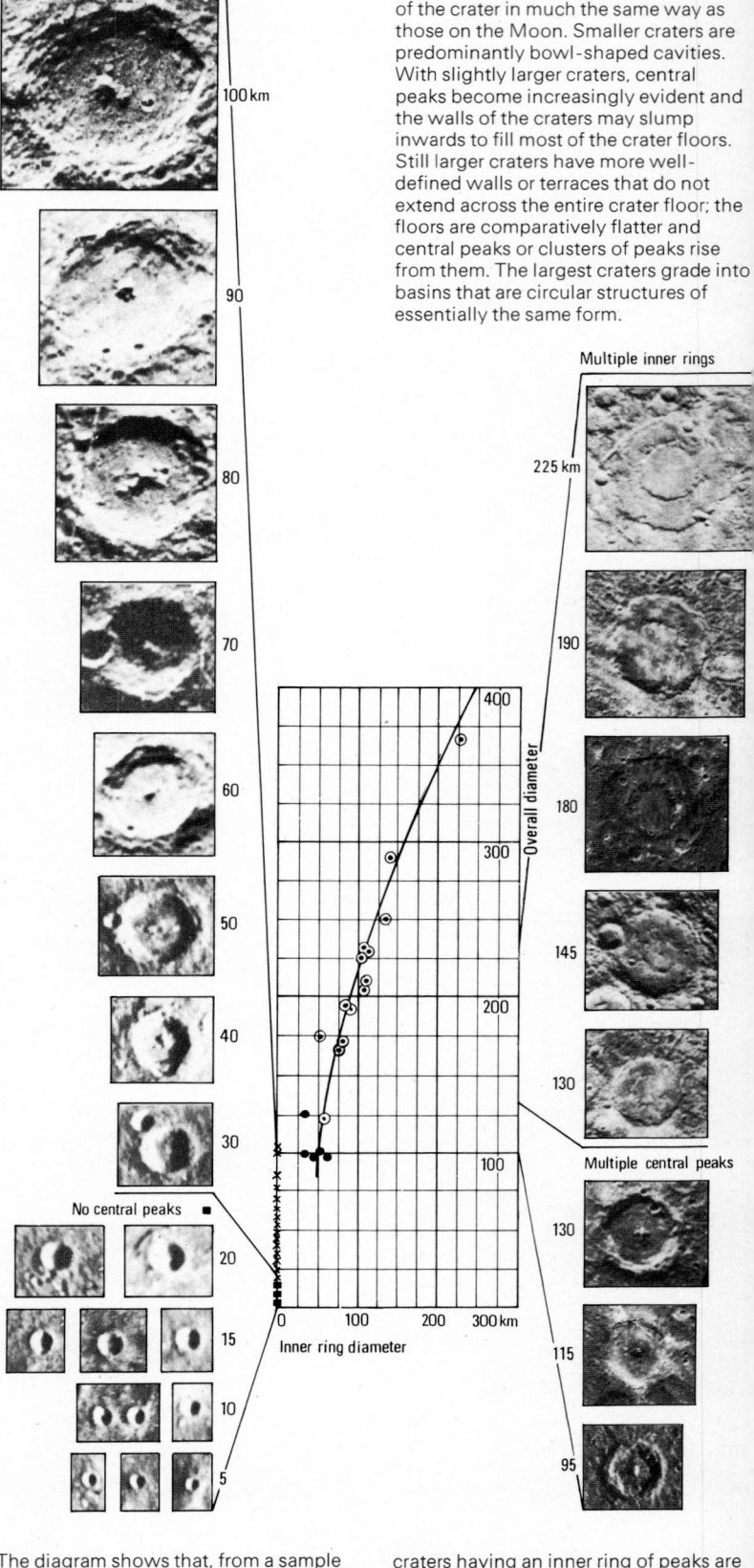

Crater types on Mercury

The interior structure of Mercurian craters is dependent on the overall size of the crater in much the same way as those on the Moon. Smaller craters are predominantly bowl-shaped cavities. With slightly larger craters, central peaks become increasingly evident and the walls of the craters may slump inwards to fill most of the crater floors. Still larger craters have more well-defined walls or terraces that do not extend across the entire crater floor; the floors are comparatively flatter and central peaks or clusters of peaks rise from them. The largest craters grade into basins that are circular structures of essentially the same form.

The diagram shows that, from a sample of Mercurian craters, the interior structure of each (defined by the inner ring diameter) is directly related to the overall diameter of the whole crater. Thus craters below 100 km in diameter are generally characterized by a central peak or a more complex structure of central elevations. Above 100 km in diameter, ringed structures of peaks, concentric with the main rim, also occur. Central peaks are absent from craters greater than approximately 130 km; above this diameter only craters having an inner ring of peaks are found. The overall sizes of the Mercurian craters are systematically less than for the craters of the Moon. Thus about 80 per cent of Mercurian craters between 10 and 20 km in diameter are terraced, and complete terracing occurs when the crater is over 20 km across. On the Moon only 12 per cent in the 10- to 20-km class are terraced and for the most part only craters greater than 40 km in diameter are consistently terraced. (In keeping with official measurements all values are given in kilometres.)

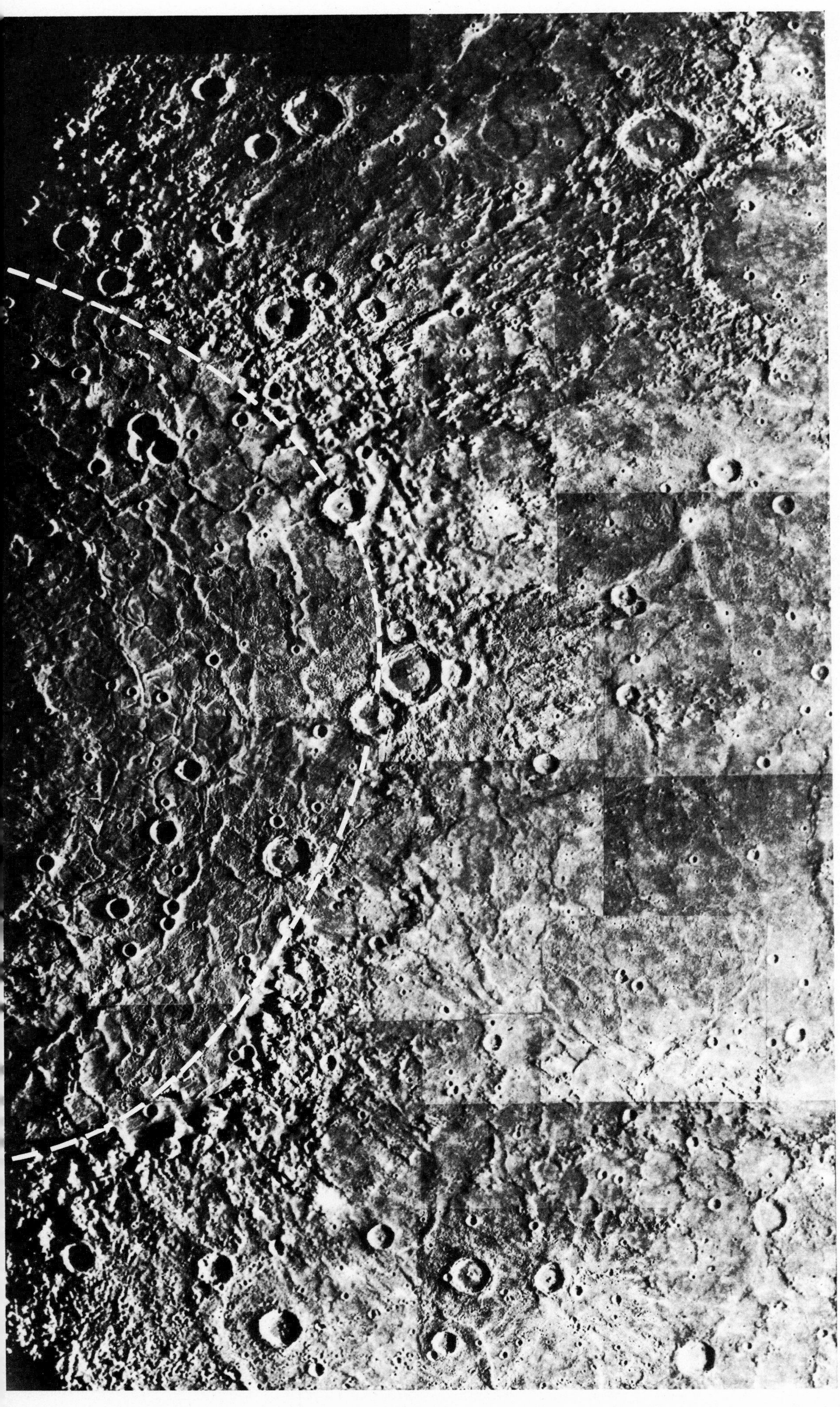

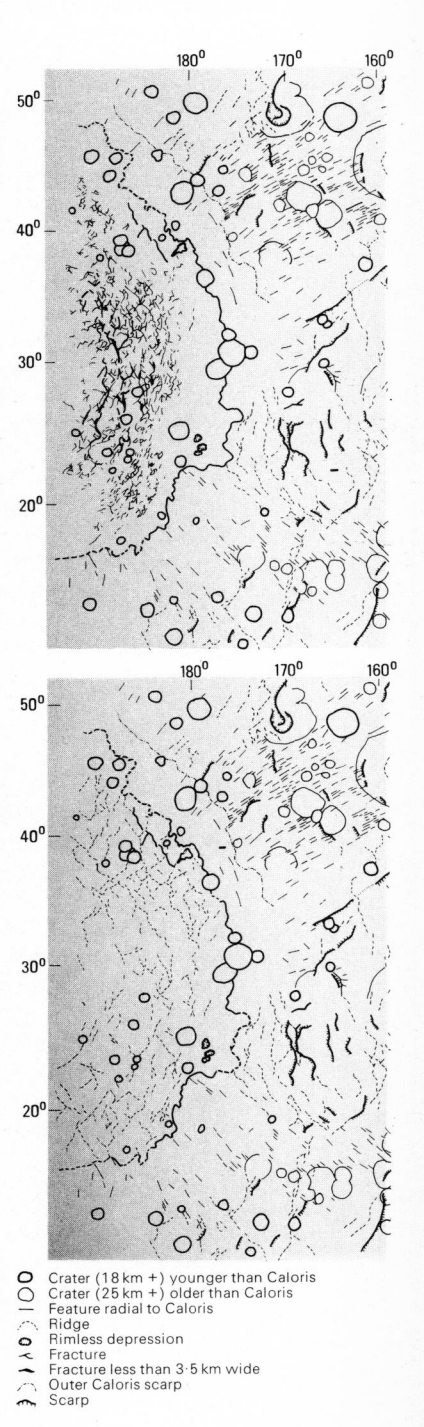

The structure of the Caloris region

The Caloris Basin (dotted) is a ringed structure—there is every reason to suppose that the part of it unavailable to Mariner 10 is as regular as the part recorded—and the effects of its formation are widely shown over the surrounding regions. The diagram, *below*, is a preliminary structural map based on the photomosaic, *left*. Fractures on the floor of the basin are much in evidence; however, information is still not complete because the angle of illumination over the basin was very low at the time of photographing.

Crater (18 km +) younger than Caloris
Crater (25 km +) older than Caloris
Feature radial to Caloris
Ridge
Rimless depression
Fracture
Fracture less than 3·5 km wide
Outer Caloris scarp
Scarp

The second part of the structural map of the Caloris Basin, *above*, displays the ridge pattern. Smooth mountain blocks rise $\frac{1}{2}$ to $1\frac{1}{4}$ miles (1 to 2 km) above the surrounding terrain, forming a ring round the 800-mile (1,300-km) diameter basin itself. Lineated terrain extends to about 620 miles (1,000 km) from the outer edge.

The atmosphere and the interior

The traces of gas detected by Mariner Although Mariner 10 had clearly established that Mercury closely resembles the Moon in surface detail and history, the results of the other experiments carried out soon revealed that the comparison could not be taken any further. The discovery that the internal constitution of Mercury is more Earth-like than any of the other inner planets and that its magnetosphere, although not exact in every detail, is comparable to a highly scaled-down version of the Earth's, was not anticipated by the mission planners. Venus was considered to be far more comparable.

Equally, despite the assertions of such telescopic observers as Antoniadi that Mercury had a discernible atmosphere, this belief was not expected to be confirmed when it was clear that the Mercurian surface was so much like the airless world of the Moon. However, the highly sensitive airglow spectrometer was able to detect positive emissions of helium as well as emissions at shorter wavelengths, which indicated the presence of very small amounts of neon and argon. Light emissions were also detected from the night hemisphere during occultation, which proved that the gases were excited by electron bombardment as well as by solar irradiation. This would mean that Mercury exhibits auroral displays in its night skies.

The strength of Mercury's magnetic field The Solar System is filled by the solar wind, made up of an ionized gas known as plasma, which is accelerated out from the Sun by the high temperature in the solar corona. Because the plasma is ionized it reacts with a magnetic field: thus the solar wind carries the solar magnetic field. In measuring the reaction of this solar wind with the magnetic field of Mercury, Mariner produced some unexpected results. The first encounter revealed the presence of a planetary magnetic field strong enough to deflect the solar wind and to form a magnetosphere—that is, the region corresponding to the planet's magnetic field.

The existence of a magnetosphere meant that the geometry of the Earth's own magnetic field could well be applied to Mercury. The magnetic field is essentially split into two regions—the magnetosphere, associated with the planet, and the magnetosheath: the two are separated by the magnetopause. The bow shock wave is the point of deflection of the solar wind from the magnetosheath. The predictions of the positions around Mercury of the bow shock and the magnetopause were completely borne out by the third encounter.

A number of experiments corroborated the original results. Thus the plasma science instruments monitored the solar wind blowing steadily outside the bow shock, and the charged particle telescopes measured the much greater energies emitted by the electrons and protons within the magnetosphere itself: these bursts of radiation would not, however, appear to constitute the equivalent of the Van Allen belts surrounding the Earth.

The liquid iron core as source of magnetic field Some part of Mercury's large iron core must be liquid for the planet to be able to generate an external magnetic field—just as this is true for the Earth (which in fact generates a far stronger field). It has only been realized in the last thirty years that a self-excited fluid dynamo in the core of a planet gives a satisfactory explanation of all the observations of its magnetic field.

Of all the planets only Jupiter and Earth were known, before Mariner 10, to have their own magnetic field. Following the Viking results, Mars has now proved to be seismologically dead with no trace of a fluid core or significant magnetic field. Venus is not so straightforward, in that it is large and dense enough to sustain a fluid core—yet Mariner could not detect a magnetic field as it flew by the planet. So far the explanation is still a matter for conjecture, and analysis of Mercury's own magnetic field will hopefully cast more light on the whole phenomenon of planetary magnetism.

Siting of the atmospheric detectors
The instruments carried by Mariner 10 to detect the existence of a Mercurian atmosphere were the occultation spectrometer (A) and the airglow spectrometer (B). Both measured light in the extreme ultraviolet portion of the spectrum. The occultation spectrometer made its measurements at the moment of entry into Mercury's shadow; sunlight passing across the limb of the planet would have to pass through an atmospheric layer—if it existed—directly into the instrument's detectors. The airglow spectrometer made a continuous scan over the illuminated hemisphere on to the dark hemisphere.

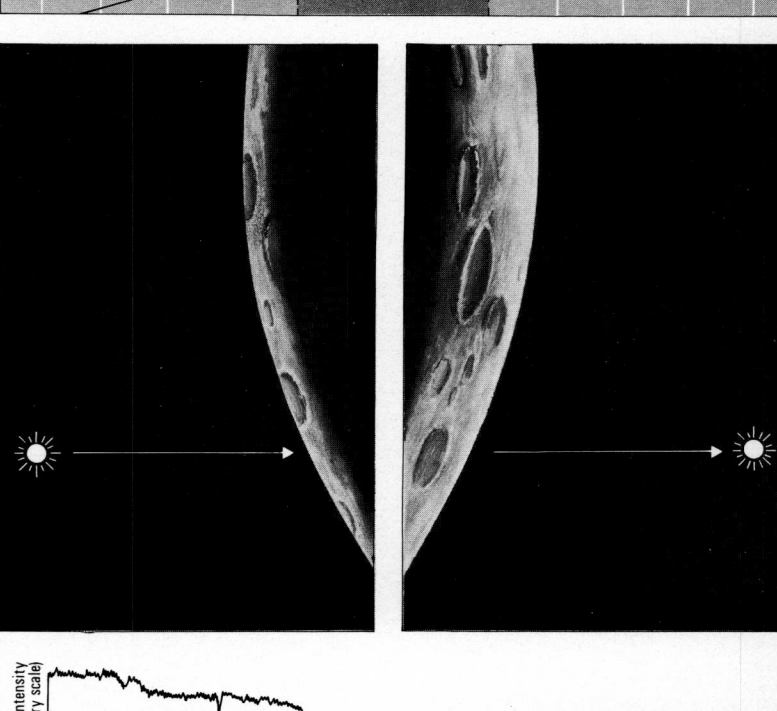

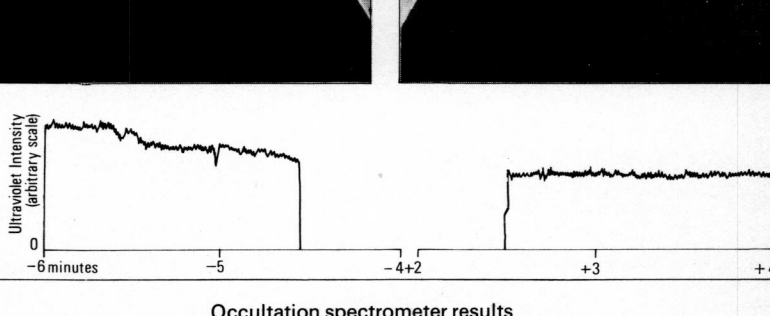

Occultation spectrometer results
Measurements could be made only during the first encounter, as this was the one time a solar occultation occurred. The approaching (*left*) and receding (*right*) limbs of Mercury are shown, with the appropriate reading for each below. The approach limb appears smaller than the receding limb as seen from Mariner, because of the vehicle's distance from the planet. The graph shows the ultraviolet light intensity is constant right up to the moment when the sunlight touched the disk of Mercury; the reading then dropped in 1·4 seconds to zero, showing that no atmospheric absorption was detected.

Airglow spectrometer results

The scanning slit (A) of the spectrometer is shown crossing the terminator of the planet. The graph below corresponds to the scan, showing the results obtained at the hydrogen emission wavelength (B) and at the helium wavelength (C). With the hydrogen line, when the slit has completed its scan, the ultraviolet intensity has returned to the level originally held, showing that no hydrogen emission has been detected. With the helium emission line, however, the intensities remain considerably higher after clearing the terminator, proving the existence of a helium airglow emission.

Mercury's magnetic field

The geometry of Mercury's magneto-sphere (A) is very similar to the Earth's, although its strength is only one-thirtieth that of the Earth's. There are two equal and opposite magnetic poles aligned with the spin axis of the planet which create the directional field shown. Seen in perspective (B) the extent of the Mariner trajectories of the first (1) and third (2) encounters is clear. Both tracks are angled towards the Sun, from which the solar wind is directed (arrowed), so that the distance leaving the magnetosphere is always shorter than the entry distance. The third encounter was planned to check

the results of the first by flying much closer to the planet with the point of closest approach near the north pole. Because it was much nearer to Mercury the field strength recorded was three times greater at maximum than that monitored by the first encounter. The measurements made as Mariner passed through the magnetic field on both encounters are shown in the graphs (C). In just over half an hour both tracks can be seen to cross the bow shock (x) and the magnetopause (y) on entering and leaving the field. These points are also located in the perspective view. It is clear that the maximum reading is at the closest approach.

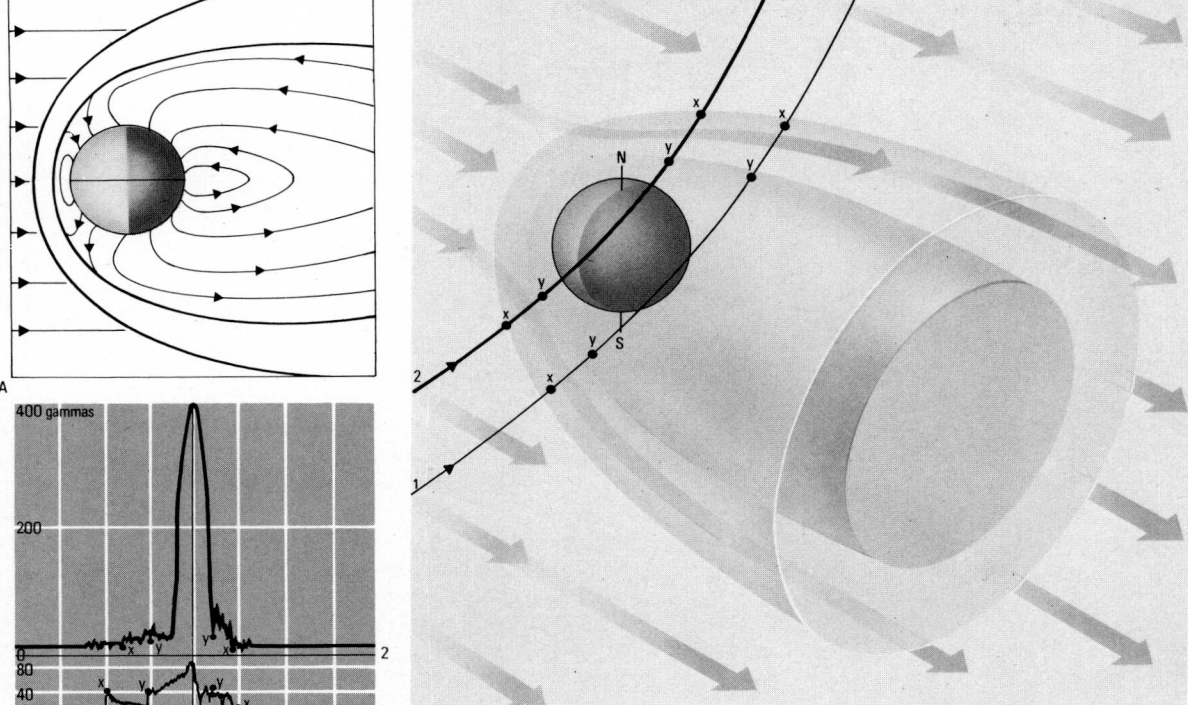

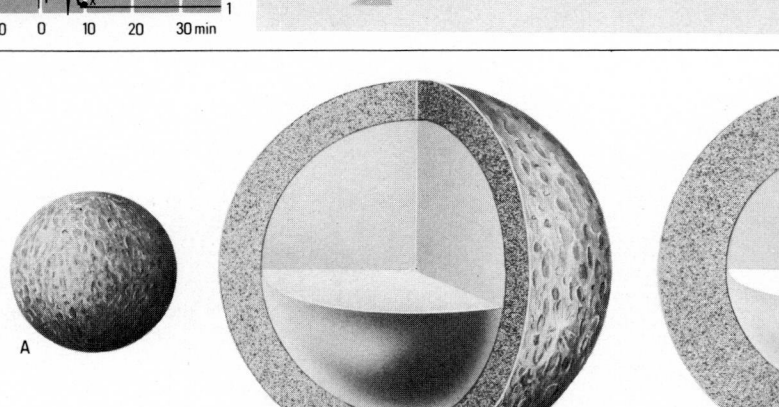

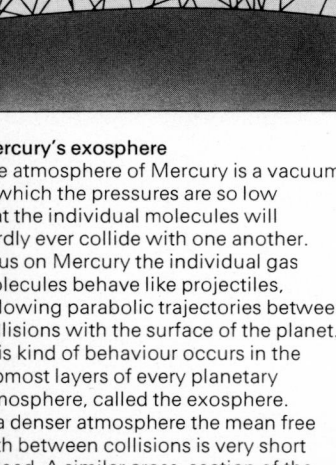

Mercury's exosphere

The atmosphere of Mercury is a vacuum, in which the pressures are so low that the individual molecules will hardly ever collide with one another. Thus on Mercury the individual gas molecules behave like projectiles, following parabolic trajectories between collisions with the surface of the planet. This kind of behaviour occurs in the topmost layers of every planetary atmosphere, called the exosphere. In a denser atmosphere the mean free path between collisions is very short indeed. A similar cross-section of the Earth's atmosphere would not show any movement of the individual molecules.

Below the Mercurian surface

Although the Earth (C) is virtually three times the size of Mercury (A) when cross-sections of the two planets are compared their internal structures are shown to bear no resemblance to their diameters. Thus Mercury's iron core (B) is comparatively much larger than the Earth's. The actual diameter of Mercury's core is estimated at 2,200 miles (3,600 km), as against 4,300 miles (6,940 km) for the Earth, which would make Mercury's core alone slightly larger than the Moon. The fact that the Earth's core is liquid with a central solid region (shown in C) has been established from detailed analyses of

earthquake waves that have passed through these regions; if they were solid, the transverse vibrations would not be able to pass through them. Mercury's magnetic field is the only evidence that its core may be similarly arranged: the Mariner results indicated that the core contains 80 per cent of the planet's mass. The fact that Mercury's core is a critical element in generating a sizeable magnetic field was not anticipated by the mission planners; it was assumed that the very slow rotation of the planet would prohibit the existence of a magnetosphere of any size. Clearly Mariner's findings will open up the study of planetary magnetism.

Glossary

Absolute zero   The lowest limit of temperature −273·16°C. This value is used as the starting point for the Kelvin scale of temperature, so that absolute zero = 0° Kelvin.

Absorption spectrum   A spectrum made up of dark lines against a bright continuous background. The dark lines, or Fraunhofer lines in the case of the Sun's spectrum, are the result of atoms absorbing radiation of the wavelength that they would normally emit.

Albedo   The reflecting power of a planet or other non-luminous body. A perfect reflector would have an albedo of 100 per cent.

Ångström unit   The unit for measuring the wavelength of light and other electromagnetic vibrations. It is equal to one hundred-millionth part of a centimetre. Visible light ranges between about 7,500Å (red) down to about 3,900 Å (violet).

Aphelion   The orbital position of a planet or other body when farthest from the Sun.

Astronomical unit   The distance between the Earth and the Sun. It is equal to 92,957,000 miles (149,596,000 km).

Aurora   Aurorae, or polar lights, are caused by bombardment of the atmosphere by energetic ions (see **Ion**) guided into polar regions by the planetary magnetic field.

Bow shock wave   That part of a planet's **magnetosphere** which deflects the **solar wind.**

Cassegrain telescope   A type of reflecting telescope (see **Reflector**) in which the light from the object under study is reflected from the main mirror to a convex secondary mirror, and thence back to the eyepiece through a hole in the main mirror.

Collimation plates   Slitted plates so arranged that a beam of parallel rays of light or other radiation is obtained.

Cusp   The point or horn of Mercury or any other planet that shows **phases**.

Density   The mass of a given substance per unit volume measured as specific gravity. The specific gravity of Mercury is 5·4—that is, unit mass of Mercury would be 5·4 times as great as that of an equal volume of water.

Ecliptic   The plane of the Earth's orbit or the apparent yearly path of the Sun against the stars, passing through the constellations of the Zodiac.

Ejecta sheet   Matter thrown out of a crater on to a surrounding surface.

Electromagnetic spectrum   The full range of what is termed electromagnetic radiation: gamma rays, X-rays, **ultraviolet**, visible light, **infra-red** and radio waves. Visible light makes up only a very small part of the whole electro-magnetic spectrum.

Electron   A fundamental particle carrying a unit negative charge of electricity.

Gibbous   A phase of Mercury or other planet that is more than half, but less than full.

Gravitation   The force of attraction that exists between all particles of matter in the universe. Particles attract each other with a force that is directly proportional to the product of their masses and inversely proportional to the square of the distance between them.

Greenwich Meridian   The line of longitude that passes through the Airy transit circle at Greenwich observatory. It is taken as longitude 0°, and is used as the standard throughout the world.

Inferior conjunction   Applied to Mercury and Venus, the position of the planets when they are on the near side of the Sun as seen from Earth.

Infra-red radiations   Radiations with wavelengths longer than that of red light (about 7,500 **Ångströms**) which cannot be seen visually. The infra-red region extends up to the short-wave end of the radio part of the **electromagnetic spectrum**.

Ion   An atom that has lost or gained one or more **electrons**; it has a corresponding positive or negative electrical charge, since in a complete atom the positive charge of the nucleus is balanced out by the combined negative charge of the electrons. The process of producing an ion is termed ionization.

Limb   The edge of the visible disk of Mercury or any other planet as seen from space.

Magnetopause   Outer limit of a planet's magnetic field and boundary of the **magnetosphere.**

Magnetosheath   Region of magnetic turbulence, outside the **magnetosphere**, in which both the magnitude and direction of the magnetic field of a planet vary erratically.

Magnetosphere   The magnetic field associated with a planet. Apart from Jupiter and the Earth, Mercury is the only known planet with a magnetosphere.

Molecule   A stable association of atoms, where they are linked together to form a group. For example, a water molecule (H_2O) is made up of two hydrogen atoms and one atom of oxygen.

Occultation   The apparent covering up of one celestial body by another. Thus as Mariner 10 flew by the dark side of Mercury, the planet occulted the Sun as viewed from the spacecraft.